Forest Entomology in West Tropical Africa:
Forests Insects of Ghana

Forest Entomology in West Tropical Africa: Forest Insects of Ghana

Michael R. Wagner
Joseph R. Cobbinah
Paul P. Bosu

 Springer

A C.I.P. Catalogue record for this book is available from the Library of Congress.

ISBN 978-1-4020-6506-4 (HB)
ISBN 978-1-4020-6508-8 (e book)

Published by Springer,
P.O. Box 17, 3300 AA Dordrecht, The Netherlands.

www.springer.com

Printed on acid-free paper

Foreword

It is indeed an honour for me to write the foreword to the second edition of *Forest Entomology in West Tropical Africa: Forest Insects of Ghana*. Originally coauthored by Professor Michael R. Wagner, Dr. S.K.N. Atuahene, and Dr. Joseph R. Cobbinah, the authorship now includes Dr. Paul P. Bosu, who replaces Dr. Atuahene of blessed memory (may his soul rest in perfect peace). In the preface to the first edition the authors stated their motivation for writing in these words: "We have written this book with the practicing forester and forest entomologist in mind." I am convinced beyond any doubt that the book has been, and continues to be, of immense benefit to the targeted audience.

Over the course of the last 16 years, since the publication of the maiden edition in 1991, there have been tremendous changes in the science and practice of forest entomology in West Africa. Research has improved significantly and the volume of literature on the subject has also increased severalfold. The biology, ecology, and management of major insects for which very little was known nearly two decades ago have now been studied. At the same time, collaboration among forest entomologists and other scientists of West Africa improved significantly and greatly enhanced the science and practice of forest entomology in the subregion. The healthy development would greatly facilitate management of pest outbreaks, which invariably are not limited by national boundaries.

The rise of invasive species appears to be one of the major challenges that humankind has to contend with in the 21st century. Deforestation coupled with increased international trade has contributed significantly to the spread of invasive species worldwide. Of these, insect pests and diseases constitute a great number, and their ecological and economic implications are huge. Since the publication of the first edition nearly two decades ago, several major pest outbreaks have been recorded in West Africa. A very good example is the outbreak of the oriental yellow scale insect *Aonidiella orientalis* which killed millions of neem (*Azadirachta indica*) trees in parts of West and Central Africa. Clearly, forest entomology as a discipline will play a major role in sustainability of West African forests over many decades to come.

In Ghana a national forest plantations development programme was launched in 2001 with a national of 20,000 hectares per annum. So far 105,000 hectares has been planted. The current plantation expansion programme could complicate pest problems in forestry and mark the beginning of outbreaks never before experienced in the country. I believe that this second edition will more than sustain the vision of the first and help in the study and management of our forests for the benefit of both the present and future generations.

Most of the information in the original edition has been updated with literature from the work of the authors and many other scientists who have worked in Ghana and West Africa. Prof Mike Wagner has been a significant player in the study and practise of forest entomology in Ghana, in particular, and West Africa in general. He has been an inspiration to countless number of Ghanaian forestry professionals, researchers and students. His association with forestry in Ghana began during his Peace Corps days as a tutor at the Sunyani Forestry Training School in the early 1970's. Together with my predecessor, Dr. Joe Cobbinah, an accomplished forest entomologist they have carried the subject beyond Ghana, and indeed beyond the shores of Africa to the international arena. Dr. Cobbinah was until recently the Co-ordinator of the International Union of Forest Research Organizations (IUFRO) Working Party on the Protection of Forest in the Tropics (WP 7-03-09). I believe that these two gentlemen have laid a good foundation and I am convinced that the next generation of forest entomologist of which Dr. Paul Bosu represent will excel. In this era of global climate change, with its associated complications on all manner of life and habitats the challenges of tomorrow's entomologist and forest health experts will without doubt be greater. I commend the authors for their foresight and efforts towards the publication of this second edition. It is a significant improvement over the first edition.

In addition to the improved line drawings and descriptions of the insects in the text, the inclusion of many good quality colour photos has added attraction to the book. This book will greatly meet the needs of forest entomologists, foresters, scientists, teachers, students and the general public.

Dr. V. K. Agyeman
Director
Forestry Research Institute of Ghana
Kumasi, Ghana
October, 2007

Preface

We believe that the second edition of *Forest Entomology in West Tropical Africa: Forest Insects of Ghana* is a significant improvement over the first edition published in 1991. This book remains the only major comprehensive effort to summarize the important forest insects of any country in West Africa. Changes to the second edition include major updating of technical information on sap-feeding insects, in particular *Phytolyma lata*. Much of the new information on *Phytolyma* comes from cooperative research activities of the authors over the last two decades. Additionally the chapter on wood borers of living trees has been updated to include major new understanding about the biology and management of mahogany shoot borer *Hypsipyla robusta*. This new information comes from a variety of worldwide sources reflecting the status of this insect as a circumglobal pest of tropical mahogany trees. We also incorporate the significant additional research on mahogany shoot borer that was conducted in West Africa since the first edition. A new chapter is added on utilitarian use of forest insects in which we discuss two topics that reflect our recognition that forest insects are pests that not only damage trees but also have potential value in direct use. Butterfly-based ecotourism is an emerging enterprise in Ghana and offers the potential of a forest-based enterprise that does not require extraction of forest products. The management of honey bees is a second example discussed in this chapter on how insects can provide direct benefit to humans but are part of the forest ecosystem.

The second edition has been written with the practicing forester and forest entomologist in mind, as was the first edition. Because the technology for reproduction has improved since the first edition we have now included a front color section. We believe that the addition of color plates makes this book more useful for practicing professionals to identify forest insects and their damage. Improvements in reproduction technology also allow us to enhance the quality of other line drawings and black and white photos used in other sections of the book.

For historical purposes we report specific chemicals that have been used in the past for insect control. Many of these chemicals may no longer be available or are inappropriate recommendations at the present time. It is important that the reader recognize that mention of any chemicals in this book does not constitute a recommendation. Pesticide recommendations change constantly and potential users of pesticide should seek current advice from local forest and agricultural officials.

The second edition is produced with sadness over the death of our beloved col-
league Dr. S.K.N. Atuahene but the joyful addition of Dr. Paul P. Bosu of Forestry
Research Institute of Ghana. We hope this book will continue to stimulate interest
in tropical forest entomology around the world. There are many fascinating forest
insects and much more scientific work needed in Ghana and all of tropical Africa.

Acknowledgments

The history of this book must be described as a long journey with many stops, starts, detours, and plenty of potholes. The real beginning was in 1974 when the senior author served as a US Peace Corps volunteer at the Forestry Training School in Sunyani, Ghana. The first draft was completed in 1975 but remained largely unused until 1985 when it was printed by the then Forest Products Research Institute as an unnumbered technical report. In 1988, with the support of a US Fulbright Senior Research Award, the project was resurrected and finally published in 1991. This second edition was initiated when Dr. Paul Bosu was a Doctoral candidate at Northern Arizona University and he expressed a willingness to join the team of authors. His persistence and energy were instrumental in getting the second edition underway.

The many individuals who contributed to the first edition were acknowledged in that volume and their names will not be repeated here. Most photographs used in the book were originals taken by the senior author. Artwork and electronic enhancement of photos was provided by the staff at the Ralph M. Bilby Research Center at Northern Arizona University. Among those contributors are Dan Boone, Ryan Belnap, and Victor Leshyk. Graduate students Monica Gaylord and Sky Stephens in forest entomology at Northern Arizona University School of Forestry kept research programs functioning and endured the frequent absence of the senior author while working on this edition. Many individuals at the Forestry Research Institute of Ghana supported this effort including: James Appiah Kwarteng, Sylvester Kundaar, Esther Amponsah, Elvis Nkrumah, Edmund Osei Owusu, Joseph Sebuka.

A very special thanks to Jennifer Stalter who word-processed, edited, created electronic images from original photos, and pretty much directed the entire logistical effort needed to complete this book. Jenn also designed the book cover.

We thank our spouses, Hathia Cobbinah, Dinah Bosu, and Karen Clancy, for creating the supportive environment we needed to see this project through to completion.

Financial support for various aspects of the production of this book was provided by Northern Arizona University School of Forestry, United States Forest Service Rocky Mountain Research Station and the International Tropical Timber Organization.

Contents

Foreword   v

Preface   vii

Acknowledgments   ix

Color Plates   xv

Chapter 1 **Forest Entomology in Ghana** **1**
            Introduction   1
            Historical Perspective   3
            Forest Insect Surveys   6
            Forest Entomology Literature   7
            The Nature of Ghana's Forests   7
            Forest Types   8
            Wet Evergreen   8
            Moist Evergreen   9
            Moist Semi-Deciduous   9
            Dry Semi-Deciduous   10
            Guinea Savannah   11
            Sudan Savannah   11
            Forest Reserves   12
            Forest Plantations   12
            Agroforestry   14
            Forest Resource Condition   21

Chapter 2 **Defoliating Insects** **23**
            Introduction   23
            Outbreaks of Defoliating Insects   24
            Types of Defoliation   24
            Lepidopterous Defoliators   24
            Anaphe venata Butler (Lepidoptera: Notodontidae)   24
            Lamprosema lateritialis Hampson (Lepidoptera: Pyralidae)   26
            Strepsicrathes rhothia Meyrick (Lepidoptera: Tortricidae)   29

Godasa sidae Fabricius (Lepidoptera: Arctiidae).   30
Epicerura pulverulenta Hampson (Lepidoptera: Notodontidae) .   31
Cirina forda Westwood (Lepidoptera: Saturniidae).   32
Miscellaneous Defoliators   32
Orthopteroid Leaf Feeder   33
Zonocerus variegatus Linnaeus (Orthoptera: Acrididae).   33
Leaf-Feeding Beetles   36

Chapter 3 **Sap-Feeding Insects** **41**
Introduction.   41
Phytolyma spp. Scott (Hemiptera: Psyllidae).   42
Integrated Control of *Phytolyma spp.*.   44
       Host Resistance.   45
       Silvicultural Control   45
       Natural Control Agents.   47
Diclidophlebia spp. (Hemiptera: Psyllidae)   47
Scale Insects (Hemiptera: Coccidae)   49
Aonidiella orientalis Newstead (Hemiptera: Diaspididae).   50
Rastrococcus invadens Williams (Hemiptera: Pseudococcidae) .   53

Chapter 4 **Wood Borers of Living Trees** **59**
Introduction.   59
Classifications of Wood Borers   59
Host Condition   59
Host Range   60
Developmental Stage of Trees When Attacked   60
Feeding or Shelter Location   60
Pest Management   61
Lepidopterous Borers of Living Trees   61
       Mahogany Shoot Borer, *Hypsipyla robusta* (Moore)
       (Lepidoptera: Pyralidae).   61
       Emire Shoot Borer, *Tridesmodes ramiculata* Warr
       (Lepidoptera: Thyrididae).   65
       Eulophonotus obesus (Lepidoptera: Cossidae).   67
       Orygmophora mediofoveata Hamps
       (Lepidoptera: Noctuidae)   68
Coleopterous Borers of Living Trees   69
       The polyphagous beetles, *Apate monachus*
       and *A. terebrans* (Coleoptera: Bostrichidae)   69
       Longhorn Borer of Bombacaceae, *Analeptes trifasciata*
       (Coleoptera: Cerambycidae).   72
       The Borer of Wawa, *Trachyostus ghanaensis*
       (Coleoptera: Platypodidae).   75
       The Ofram Borer, *Doliopygus dubius* Samps
       (Coleoptera: Platypodidae).   77
       Hypothenemus pusillus (Coleoptera: Scolytidae).   79

Chapter 5   Pests of Flowers, Fruits, and Seeds **87**
   Introduction.   87
   Guarea Fruit Weevil, *Menechmaus sp.* (Coleoptera:
   Curculionidae)   88
   Fruit borer on wawa, *Apion ghanaensis* and *A. nithonomiodes*
   (Coleoptera: Apionidae)   90
   Emire seed weevil, *Nanophyes sp.* (Coleoptera: Curculionidae) .   93

Chapter 6   Pests of Logs, Lumber, and Forest Products **101**
   Introduction.   101
   Ambrosia Beetles   102
   Scolytidae: Taxonomy and Biology   102
   A Typical Scolytid Beetle: *Xyleborus ferrugineus*   103
   Platypodidae: Taxonomy and Biology   104
   A Typical Platypodid Beetle; *Doliopygus conradti*   104
   Ambrosia Beetle Damage (Scolytidae, Platypodidae)   105
   Phloem Borers   108
   Longhorn Beetles (Cerambycidae).   108
   Metallic or Flatheaded Wood Borers (Buprestidae).   125
   Powderpost Beetle   127
   Bostrichidae   127
   Lyctidae.   134
   Freshwater Wood Borer   135
   Taxonomy and Biology.   135
   Miscellaneous Wood Borers.   139

Chapter 7   Termites **147**
   Introduction.   147
   Termite Castes.   147
   Caste Determination   148
   Colony Formation.   149
   Identification.   150
   Drywood Termites   152
   Subterranean Termites   155
   Termites Attacking Living Trees   161
   Beneficial Role of Termites   163

Chapter 8   Utilitarian Use of Forest Insects **167**
   Introduction.   167
   Butterfly-Based Ecotourism.   167
      Ecotourism Defined   168
      Ecotourism and Economic Development.   168
      Why Butterflies?   168
      Bobiri Butterfly Sanctuary.   169
   Apiculture.   170

Beekeeping with African Bees in the Tropics 171
Beehives. 171
Stingless Bees (Other Bees; Sweat Bees) 172
Production and Marketing of Honey in Ghana 172

References. **173**

Appendix A **183**

Appendix B **207**
Termite Species Recorded in Ghana 207

Appendix C **209**
Non-Native Trees that have been Introduced into Ghana 209

Appendix D **213**

Glossary **227**

Index to Insect Scientific Names **239**

Color Plates

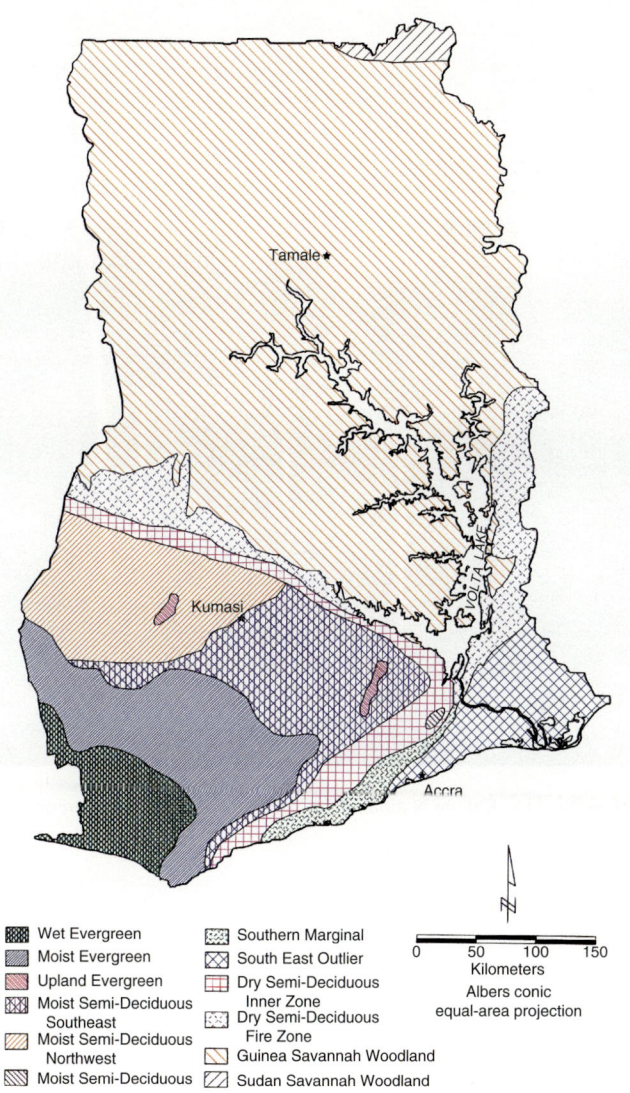

Wet Evergreen
Moist Evergreen
Upland Evergreen
Moist Semi-Deciduous Southeast
Moist Semi-Deciduous Northwest
Moist Semi-Deciduous

Southern Marginal
South East Outlier
Dry Semi-Deciduous Inner Zone
Dry Semi-Deciduous Fire Zone
Guinea Savannah Woodland
Sudan Savannah Woodland

0   50   100   150
Kilometers
Albers conic
equal-area projection

Plate 1 Forest vegetations cover types in Ghana. Adapted from a map provided by Ghana Forest Services Division, Kumasi.

Plate 2 Wet evergreen forest at Kakum National Park. High species diversity and multiple canopies are typical.

Plate 3 Moist Semi-Deciduous forest at Bobiri Forest Reserve. Prominent tree on left is *Khaya ivorensis.*

Plate 4 *Theobroma cacao* is a major agricultural crop and the source of cocoa powder used as the key ingredient in chocolate.

Plate 5 Dry Semi-Deciduous forest cover type near Abofour, Ghana. Emergent tree in center of photo is *Milicia excelsa*.

Plate 6 Guinea savannah at Mole National Park, Ghana.

Plate 7 Sudan savannah near Bolgatanga, Ghana. Subsistence farming is common in the savannah region.

Plate 8 Baobob, *Adansonia digitata*, is one species that typifies the Sudan savannah. Fruits and foliage of this species are used as human food.

Plate 9 *Faidherbia albida* (right side of photo) is common in the Sudan savannah and has asynchrony in foliage production. Leaves are present during the dry season and absent during the rainy season.

Plate 10 Teak, *Tectona grandis*, is the most commonly planted exotic species for plantations in Ghana and throughout West Africa. Teak plantations are fire resistant and local farmers commonly burn them for weed control.

Plate 11 Plantation of native *Terminalia superba*. Some *Terminalia* species produce allelopathic compounds that limit their use in pure plantations.

Plate 12 Experimental mixed plantation at the Mesewan nursery near Kumasi.

Plate 13 Dr. Francis Ulzen-Appiah from Kwame Nkrumah University of Science and Technology. *Cedrela odorata* (center) is a fast growing exotic species now widely planted in mixed plantation and agroforestry systems.

Plate 14   Demonstration of agroforestry practice of alley cropping at Kwame Nkrumah University of Science and Technology.

Plate 15 Snails, *Achatina achatina*, are an important non-timber forest product throughout West Africa.

Plate 16 *Lamprosema lateritialis* layered egg mass on mid-rib of *Pericopsis elata* leaves.

Plate 17   Early stage in the development of a compound nest of *Lamprosema lateritialis* made by tying leaves together with silk.

Plate 18   Fully developed nest of *Lamprosema lateritialis* with evidence of near complete webbing of leaves together and extensive defoliation.

Plate 19 *Pericopsis elata* seedlings heavily defoliated by *Lamprosema lateritialis.*

Plate 20 *Lamprosema lateritialis* late instar larva.

Plate 21   *Cirina forda* (Lepidoptera: Saturniidae) defoliating shea nut tree (*Vitellaria paradoxa*) in northern Ghana.

Plate 22   Defoliated branch of *Vitellaria paradoxa*.

Plate 23 Traditional production of shea butter from shea nuts. Shea nuts are scattered in the foreground.

Plate 24 Leaf of teak (*Tectona grandis*) completely skeletonized by grasshoppers (*Zonocerus variegatus*).

Plate 25 Pair of *Zonocerus variegatus* grasshoppers.

Plate 26   *Phytolyma lata* (Hemiptera: Psyllidae), is a major sap sucking pest on *Milicia* spp.

Plate 27 Galls and dieback on *Milicia excelsa* seedling caused by *Phytolyma lata*. Colonization of galls by saprophytic fungi causes the characteristic dieback.

Plate 28   Large *Phytolyma* gall mass on *Milicia* spp. some tissue is cut away to expose chambers within the gall where nymphs reside.

Plate 29   *Phytolyma* gall that has burst open releasing psyllid inside. Gall bursting is essential to release psyllid from inside because psyllids lack chewing mouth parts.

Plate 30 *Milicia* seedling that has lost its leaves exposing extensive number of large galls.

Plate 31 Dr. Paul Bosu standing in provenance plantation of *Milicia* spp. Note high variability of diameter of saplings related to genetic resistance to *Phytolyma*.

Plate 32 *Phytolyma* galls that are from highly susceptible genotypes (left two) versus galls that are from resistant or tolerant genotypes (right two). Small hard galls on right do not burst open to release adult psyllid.

Plate 33 Professor Mike Wagner (left) and Dr. Joe Cobbinah (right) next to a 7 year old highly tolerant *Milicia* genotype that grew quickly despite heavy populations of psyllids in the nursery.

Plate 34 *Milicia excelsa* growing in the shade of *Gliricidia sepium*. *Milicia* damage by *Phytolyma* is reduced when grown under shade as compared to when grown in full sun.

Plate 35   Dr. Paul Bosu in an experimental mixed plantation. Maintaining *Milicia* in low density in a mixed population is an effective control strategy currently under development.

Plate 36 Vegetation propagation of *Milicia* genotypes tolerant of *Phytolyma*.

Plate 37   Tissue culture produced seedlings of genetically resistant/tolerant *Milicia* seedlings.

Plate 38   Large, high value *Milicia* spp. log from primary forest harvest in Cote d' Ivoire.

Plate 39 *Triplochiton scleroxylon* (wawa) sapling attacked by the psyllid *Diclidophlebia* spp. (Hemiptera: Psyllidae).

Plate 40 Curling and discoloration caused by *Diclidophlebia* spp. of wawa leaf (left) as compared to a healthy leaf (right).

Plate 41 Severely damaged wawa leaf including chlorosis and marginal leaf necrosis following attack by *Diclidophlebia* spp.

Plate 42 Mango (*Mangifera indica*) leaves heavily infested with mango mealy bug, *Rastrococcus invadens*.

Plate 43   Close up of mango mealy bugs.

Plate 44 Feeding damage to bark tissue of *Khaya* spp. by the mahogany shoot borer *Hypsipyla robusta.*

Plate 45 Feeding damage to *Khaya* spp. shoots by the mahogany shoot borer.

Plate 46 Exposed larva of mahogany shoot borer.

Plate 47 Multiple stems caused by mahogany shoot borer attack.

Plate 48 Copious resin production in *Khaya* sp. in response to attack by mahogany shoot borer.

Plate 49   Baobob, *Adansonia digitata*, with numerous witches brooms caused by repeated attacks from *Analeptes trifasciata*.

Plate 50   Pair of *Analeptes trifasciata* adults on baobob.

Plate 51 Stem girdle caused by *Analeptes trifasciata*. Adults girdle the stem then lay their eggs on the distal portion of the girdled branch.

Plate 52 Gallery in wawa caused by *Trachyostus ghanaensis*.

Plate 53 Boring dust on surface of log caused by ambrosia beetles.

Plate 54 Typical blue/black stain associated with ambrosia beetle galleries in wawa.

Plate 55 Standard practices in tropical forest harvesting are to peel logs shortly after felling to avoid wood borer attack.

Plate 56 Surface view of powder post beetle damage to wawa.

Plate 57 Treating lumber by dipping into tank of wood preservative. Common procedure before export.

Plate 58 Subterranean termite mound in Guinea savannah near Buipe, Ghana.

Plate 59 Fungus gardens within a subterranean termite mound.

Plate 60 Close-up of convoluted wood fiber mat upon which fungi are cultivated by termites.

Plate 61 Termite queen encased in a hard clay "royal chamber".

Plate 62 Fruiting bodies of the termite cultivated fungus, *Termitomyces* sp. The fruiting bodies will open to a typical mushroom. They are edible and quite delicious.

Plate 63 Standard tropical butterfly live trap usually baited with fermenting banana. Butterflies can be identified and released from trap unharmed.

Plate 64 Butterfly feeding station. Butterflies are attracted to fermenting fruits.

Plate 65 Gardens established to attract butterflies for public viewing. Bobiri Forest Reserve and Butterfly Sanctuary.

Plate 66 Large red flowers of *Clerodendron* sp. that is highly attractive to butterflies.

Plate 67 Bobiri Forest Reserve and Butterfly Sanctuary Guest House within the Bobiri Forest Reserve, Kubease, Ghana.

Plate 68 Butterflies painted by local artists direct visitors to Bobiri Butterfly Sanctuary reception area.

Plate 69   Butterfly biology interpretive sign.

Plate 70 Forest trails within Bobiri Butterfly Sanctuary that visitors use to enjoy the natural forest.

Plate 71 Tropical butterfly, *Salamis cyctora* at Bobiri Butterfly Sanctuary. Identification courtesy of Dr. Janice Bossart.

Plate 72 *Junonia sophia* butterfly at Bobiri. Identification courtesy of Dr. Janice Bossart.

Plate 73 *Euphaedra janetta* butterfly. Tentative identification courtesy of Dr. Janice Bossart.

Plate 74 Kenya top bar hive system used for honey reproduction with European bees in Ghana.

Plate 75 Beekeeping in Ghana using Kenya top bar hives. Note the formation of combs on the top bar.

Chapter 1
Forest Entomology in Ghana

Introduction

Forest entomology is the study of insects (entomology) as it relates to the management of forestlands to meet the goals of stakeholders. This field has its roots in entomology and forestry; most forest entomologists have training in both of these disciplines. The subject "forest entomology" is most often taught in forestry schools because foresters have recognized the potential importance insects can play in meeting their management objectives for the forestland under their care. Because of the complex nature of forestry, it is not possible for students of forestry to obtain the full depth of training in entomology that is necessary to manage all forest insects under all conditions. It is important, however, that forestry students are aware of the importance of forest insects and learn to recognize damage when it occurs.

It is difficult to assign a specific monetary value to the damage insects cause to forests and forest products. It is, however, safe to say that all forest species in Ghana have at least some insect pests that can cause serious damage in some situations. It is also generally true that the economic importance of forest insect damage is greater after trees have been processed into wood products. Applying control to mitigate insect damage in a natural forest may be difficult to justify when many years may remain before harvesting occurs. However, a pest problem in processed lumber scheduled for export or in a nursery might easily justify considerable expenditure of funds to control. For example, Atuahene (1970) noted that dipping lumber in an insecticide to control insect wood borers was essential to meet export standards. The cost of treatment alone was 20,000 cedis in 1968. In 1968, 20,000 cedis (2.00 Ghana cedis) would pay the salary of a technical officer for many years. Specific monetary sums lose their meaning in inflationary times, but a general rule of thumb is that insects get about as much of the forest product as we do. Considering current estimates (FAO 2005), that puts wood loss to insects at around 23 million m^3/year. In Ghana's natural resource-based economy, that is a significant figure indeed!

Though many insects have the potential to devastate forests or their products, not all insects are damaging! In entomology we use the term "pest" or "pest insects" to distinguish insects which damage our food and fiber from innocuous or beneficial species. The term "pest" is anthropocentric in its origin. If an insect interferes with a specific management objective, it is considered a pest. The same insect species can be a pest in one situation and beneficial in another. For example, termites play an essential ecological role in nutrient recycling (see Chapter 7) and are a prized food source for those people who have learned how to enjoy them. In this context, termites are beneficial. However, when these same insects feed on wooden structural members in our dwellings and factories, they can be serious pests. Probably less than 5% of the total insect fauna of Ghana ever become pests. Most insects are of little importance to production of food and fiber. Another large portion of insects are beneficial. These species are predators and parasites of other insects and help keep pest insects in check. On a global basis the value of insects for positive economic value is recognized. In the last chapter we discuss insect-based ecotourism that is getting a foothold in Ghana and contributing to its local economy. The concept of a pest is important for foresters to understand so they can appropriately focus their attention on those insect species that really interfere with our forest management objectives.

An important question that can be raised is: "How is forest entomology in Ghana different from forest entomology elsewhere in the world?" Clearly, forest entomology is well developed in North America and Europe where several textbooks have been published (e.g. Barbosa and Wagner 1989; Berryman 1986; Coulson and Witter 1984; Speight and Wainhouse 1989). There are also edited volumes such as that by Watt et al. (1997) and compendia of host trees and their insect pests such as that by Browne (1968). Dajoz (2000) discussed tropical rain forest insects as part of the broader topic of forest entomology. This text, however, focuses on ecological roles of insects in tropical forest rather than traditional forest pest management. Speight and Wylie (2001) were the first to truly view tropical forest insect pest management as a subset of the historical vision of forest pest management as it had emerged in temperate regions. Serious students of tropical forest entomology should consider Speight and Wylie as the benchmark reference to begin their study. Finally, regionally focused discussions of forest entomology are emerging that begin the process of bringing forest entomology knowledge to the level of practicing foresters. Two examples for Africa are Schabel (2006) and this volume. However, these references are not sufficient for use in Ghana for several reasons, including the diverse and complex nature of tropical forests, the varied insect fauna, and the developing nation status that creates significant economic limitations on the management approaches available. This book is more oriented toward practicing foresters in West Africa who need to identify forest pests and manage them with their limited economic capability. We also discuss forest insect surveys and efforts to describe the Ghanaian insect fauna. Much of this book is devoted to consolidating information on forest insects in Ghana. Later in this chapter we will discuss briefly the nature of forests in Ghana.

The developing nation status of Ghana creates many complex problems for the development of forest entomology. Libraries and literature resources are inadequate, field and laboratory equipment is nearly nonexistent, trained personnel are few in number, economic resources are limited, and management strategies are restricted to labor-intensive approaches. In this regard little has changed since Wagner et al. (1991). Though these limitations are real, progress is being made through dedicated, hardworking staff. There are many unique insect species that should attract the attention of entomologists from around the world. What makes forest entomology in Ghana unique is that there are many more species of trees and insects than in the temperate forests, and as a consequence, there are many more complex ecological relationships. These complex ecological relationships make for difficult insect problems in an environment of great economic limitations. Many challenges exist for forest entomologists in Ghana.

Historical Perspective

Prior to independence in 1957, Ghana (Gold Coast) was part of the British Commonwealth. The timber trade began in 1891 and initially was focused primarily on the mahoganies, namely, *Khaya* and *Entandrophragma* spp. (Taylor 1960). By 1913, the timber trade was well developed in Ghana with exports greater than 84,000 m³. By 1948 timber exports had reached 196,000 m³ and peaked in 1975 at 1,300,000 m³ (Taylor 1960; Hall and Swaine 1981). Based on the importance of the timber trade in Ghana, it is not surprising that early control efforts were focused on ambrosia beetles, which are major pests of cut logs. In 1935, the need to control infestation by ambrosia beetles was identified by the Fourth Imperial Entomological Conference (Kudler 1978; Anonymous 1957). This ultimately led to the formation of the West Africa Timber Borer Research Unit (WATBRU) in 1953 in Kumasi. WATBRU was the first major research effort in forest entomology in Ghana. At about the same time, research work on termites was begun by the Termite Research Unit, Commonwealth Institute of Entomology, UK (Harris 1964). These two research units represented the major research efforts in forest entomology during the colonial period, when important field research facilities were established within the Bobiri Forest Reserve. A field research facility (Figure 1.1) and an overnight guesthouse (Figure 1.2) were built *c.*1939. Those facilities are still standing and are part of the Bobiri Forest Guest House and Butterfly Sanctuary. A great deal of the early field activities of WATBRU and related projects were done at these facilities. The current use of Bobiri as an ecotourism site is discussed in the last chapter.

The most significant event in forest entomology since independence was the establishment of the Forest Products Research Institute (FPRI) in 1964. Within FPRI a Protection Division was established which included the Entomology Section and Pathology Section. The Entomology Section took over the old facilities of WATBRU and moved from the city center to the campus of the now Kwame Nkrumah University of Science and Technology (KNUST) (Figure 1.3). The name Forest Products Research Institute was somewhat of a misnomer because their research charge extended beyond products entomology to include all classic areas in forest entomology. It is only since 1964 that research has

Figure 1.1 Original research laboratory at Bobiri Forest Reserve and Butterfly Sanctuary in need of renovation. Note butterfly attraction experiment in the foreground

Figure 1.2 Old wooden guesthouse at Bobiri Forest Reserve and Butterfly Sanctuary. Structure was constructed *c*.1940 with termite- and wood decay-resistant tropical trees

Figure 1.3 The Entomology and Pathology Sections of the Forest Products Research Institute on the campus of the University of Science and Technology, Kumasi

been conducted on general areas of forest entomology. In 1991, FPRI was changed to the Forestry Research Institute of Ghana (FORIG), which reflects the widening scope of its research activities, and was placed under the umbrella of the Council for Scientific and Industrial Research (CSIR). The work of Termites Research Unit is continuing under the direction of the Buildings and Roads Research Institute (BRRI), which was also located on the university campus. Today, both FORIG and BRRI have relocated to permanent facilities at Fumesua, about 3 km away from the university, on the main Kumasi-Accra trunk road (Figure 1.4). For a long time, FORIG (FPRI) provided the forest entomology expertise for the Faculty of Renewable Natural Resources (Institute of Renewable Natural Resources) at KNUST. FRNR provides the professional training for all foresters in Ghana.

Figure 1.4 Forest Biology building at the Forestry Research Institute of Ghana at Fumesua, near Kumasi. Both the Entomology and Pathology Sections are located in this building

Forest Insect Surveys

There has never been a systematic survey conducted of the forest insects of Ghana. This would indeed be an immense task requiring many years to complete. There have, however, been partial surveys conducted that provide useful information. Taylor (1960), in his classic work, provides some anecdotal information on insects. Thompson (1963) spent considerable time while he was Assistant Conservator of Forests from 1945 to 1949 collecting wood-boring insects, which resulted in the first important forest insect survey information in Ghana. Considerable survey-type information is available in the research reports of the WATBRU that include: WATBRU Reports 1957, 1959, 1960, 1961, 1962 (Anonymous 1957, 1959, 1960, 1961, 1962) and three Technical Bulletins (Jones 1959a; Roberts 1969). Detailed information on the biology of the important bark and ambrosia beetles was provided by Browne (1963). This work was later expanded and updated to cover a large part of the British Commonwealth (Browne 1968). At about the same time a survey of important forest insects in Nigeria was also published (Roberts 1969). A survey of agricultural insects that includes some references to forest insects that are found in Ghana has also been published (Forsyth 1966).

Annual forest insect surveys are not conducted in Ghana. However, very useful information is reported in the annual Forest Entomology Section Reports of FPRI. These reports are available for 1964, 1965, 1966, 1968, 1969/70, 1970/71, 1971/72, 1972/73, 1975/76, 1976/77, 1977/78, 1978/79, 1979/80, 1982, 1984, 1985, 1986, and 1988–2005. These reports list the forest insect problems that were observed in those years. Though these reports do not provide systematic estimates of insect populations or forest damage, they do indicate what insect problems were investigated in those years. Because many of the species determinations given in those reports were made by the British Museum, they are a reliable source of information and were used heavily in this book. A recent review indicates that insect biosystematic services are greatly needed throughout Africa, including Ghana (Ritchie 1987). A few specialized surveys have been conducted, such as the survey of insect pests affecting the Apirade nursery (Cobbinah 1972a, b). A comprehensive summary of the butterfly fauna of West Africa has recently been published (Larsen 2005).

Forest Entomology Literature

One of the many factors that make work in forest entomology difficult in Ghana is the lack of good literature. In the previous section we discussed some of the early forest entomology literature. Many of these were published outside Ghana, which limits their usefulness to Ghanaian foresters and forest entomologists. The main sources of forest entomology literature within Ghana are the FPRI Technical Newsletters and Bulletins. These reports have been published on a fairly regular basis since 1964 and contain very valuable information. The *Ghana Journal of Forestry* (formerly, *Ghana Forestry Journal*) and *Ghana Journal of Science* are locally produced scientific journals published at irregular intervals that also contain information on forest insects. There are also unpublished internal reports that include information on forest insects. We have attempted to limit our use of these unpublished reports to a minimum, but in several cases they were the only source of information available to us. Gradually we are beginning to find peer-reviewed literature on forest insects of Ghana. Much of this literature is cited in this revised edition in appropriate sections.

The Nature of Ghana's Forests

Much of the natural vegetation in Ghana is either forest or savannah; these cover types support numerous woody tree species (Plate 1). The exception is the coastal areas around Accra where grassland, coastal thicket, and mangrove vegetation exist. Roughly 55% of Ghana is the savannah type, while one third is closed forest (Hall and Swaine 1981). Because various tree species exist naturally in all of these areas, forest insect problems can be found throughout Ghana.

We accept the basic forest vegetation classification originally described by Hall and Swaine (1976) and described in detail in Hall and Swaine (1981). This classification is currently in use by the Forest Services Division. Plate 1 is reproduced from maps and includes the Hall and Swaine classification system when appropriate. We also use the term "high forests" to refer to closed forests or "forests" as described by

UNESCO (1973). "High forest" is a common term used in Ghana and is known to all foresters. The term "savannah" is used here to mean both Guinea and Sudan savannah. One of the difficulties in Ghana that we have previously mentioned is the low availability of good reference material. The Hall and Swaine (1981) text on forest vegetation in Ghana cost the equivalent of 6 months' gross pay for a Senior Technical Officer in the Forestry Department in 1988 and is currently out of print. Consequently, this most basic reference is unavailable to the practicing forester. To alleviate this problem, we briefly summarize the important characteristics of the major forest vegetation types in Ghana.

Forest Types

Wet Evergreen

This forest type represents the heaviest area of rainfall in Ghana and is the only true "rainforest." This community is floristically the most diverse in Ghana, but contains only a few of the important commercial timber species. The stands are multistoried and the soils are nutrient-poor (Plate 2). This type represents approximately 3.3% of the total forest area in Ghana (Table 1.1). Tree height is the lowest among the four major forest types (Table 1.2). Lower average tree height may be related to high rainfall and associated nutrient leaching. Degree of deciduous habitat is lowest among the forest types but can be highly asynchronous both between species and within species. Tree species that are common in the wet evergreen forest include *Dacryodes klaineana, Strombosia glaucescens, Diospyros sauza-minaka*, and *Dialium aubrevillei*. Important timber species in the wet evergreen forest include *Heritiera utilis, Lophira alata*, and *S. glaucescens*.

Table 1.1 Annual precipitation and land area occupied by the major forest and woody vegetation forest types in Ghana[1]

Forest types	Rainfall (mm)	Area (ha)	Percentage
Closed forest		2,191,910	34.6
Wet evergreen	1750–2000	209,055	3.3
Moist evergreen	1500–1800	519,470	8.2
Moist semi-deciduous	1200–1800	893,235	14.1
Dry semi-deciduous	1250–1500	570,150	9.0
Woodland		3,553,935	56.1
Guinea savannah	900–1500	3,503,255	55.3
Sudan savannah	600–900	50,680	0.8

[1] Area based on FAO (2005); percentage based on Forestry Department maps. These data have not been adjusted for potential differentials in forest loss among forest cover types.

Table 1.2 General characteristics of closed forest types in Ghana. (Summarized from Hall and Swaine 1981.)

Forest types	Tree height X/range	Commercial timber value	Floristics	Degree of deciduous habitat
Wet evergreen	X = 32 m	Lowest	Highest diversity, most characteristic species	Lowest
Moist evergreen	X = 43 m	Second to moist semi-deciduous	Lower diversity, than wet evergreen, more characteristic species than most semi-deciduous	Intermediate, asynchronous
Moist semi-deciduous	50–60 m	Highest	Many species common to all other forest types	Intermediate, synchronous
Dry semi-deciduous	30–45 m	Intermediate	Many common species	Highest, synchronous

Moist Evergreen

This type is similar to the wet evergreen type in rainfall and floristic areas (Plate 2). However, there tends to be a few more deciduous species and considerably more economic species associated with the moist evergreen type. One of Ghana's most important commercial species, *Triplochiton scleroxylon*, occurs in this type (Figure 1.5). The moist evergreen type represents approximately 8.2% of the total forest area (Table 1.1). Tree heights tend to be lower than the moist semi-deciduous trees but higher than the wet evergreen forest.

Moist Semi-Deciduous

The moist semi-deciduous forest is the most extensive closed canopy forest type in Ghana (14.1%) (Table 1.1, Plate 3). This type or quite similar types are also quite abundant throughout West Africa (Hall and Swaine 1981). Trees tend to be 50–60 m and are the tallest of any forest type. Deciduous and evergreen species are represented in about equal proportion. These are often several canopy layers and the upper canopy is discontinuous due to the presence of a few emergent species. This is the major timber-producing area, and includes species such as *Celtis mildbraedii, Entandrophragma utile, Guibourtia ehie, Khaya anthotheca, Khaya ivorensis, Nesogordonia papaverifera, Pericopsis elata, Terminalia ivorensis*, and *T. scleroxylon*. As the name implies, the moist evergreen type consists of approximately equal proportions of evergreen and deciduous species. This type is typified by tall tree species, and multistoried canopies (Plate 3), and is suitable for most forest crops including cacao (*Theobroma cacao*) (Plate 4). Hall and Swaine (1981) subdivided this type into the North-West and South-East subtypes.

Figure 1.5 *Triplochiton scleroxylon* (wawa or obeche) is one of Ghana's most economically important species and occurs in several forest types, including the moist evergreen type. This species has an emergent crown that often towers above the canopy of other trees

Dry Semi-Deciduous

This forest type occupies a range of environmental conditions and constitutes the transition zone between the higher rainfall types and the Guinea savannah. The dry semi-deciduous covers approximately 26% of the forest area. Trees achieve intermediate height compared to the other forest cover types. Vertical crown structure is well developed with multiple canopy layers. Important commercial species that occur here include *Afzelia africana, Antiaris africana, Celtis zenkeri, Cola gigantea, Milicia* spp. (Figure 1.6, Plate 5), *Sterculia tragacantha*, and *T. scleroxylon*. Hall and Swaine (1981) subdivide this forest into the Fire Zone and Inner Zone subtypes.

Figure 1.6 Dry semi-deciduous forest type near Sunyani in the Brong-Ahafo Region

Guinea Savannah

The Guinea savannah consists of tall grasses growing between widely spaced trees (usually without overlapping crowns) (Plate 6). There is, however, considerable variation in crown covers and many areas may actually qualify as woodland according to the definitions of UNESCO (1973). In much of the savannah, the two rainy seasons typical of southern Ghana become one rainy season and one dry season (Lawson 1986). Most of the tree species growing on the Guinea savannah are deciduous. The soils are quite fertile and capable of supporting a variety of crops. The Guinea savannah is the most extensive forest type in Ghana (55.3%), but contains few commercial tree species. However, teak (*Tectona grandis*) plantations grow well and are of commercial value. Some important tree species in this type include *Lophira lanceolata, A. africana, Parkia clappertoniana, Daniellia oliveri*, and the important oil-producing species, sheanut tree (*Vitellaria [Butyrospermum] paradoxa*) (Lawson 1986).

Sudan Savannah

This region lies in the northeastern part of Ghana and occupies a relatively small total area (Table 1.1, Plate 7). The Sudan savannah has two tree species that are quite characteristic of the type, *Faidherbia [Acacia] albida* and baobob, *Adansonia digitata* (Plate 8). *F. albida* has the unusual habit of dropping its leaves during the rainy season and refoliating during the dry season (Plate 9). Because of this characteristic,

F. albida is widely recommended as an agroforestry species. The baobob is one of the most interesting and picturesque trees in the savannah. Baobob foliage and fruits are utilized by local people as food.

Forest Reserves

The Forestry Department (now Forest Services Division) was established in Ghana in 1909, but it was not until 1927 that the Ghanaian government had the power to set aside land to prevent widespread destruction of the forest. Forest reserves were established between 1927 and the late 1970s; no new reserves have been established since that time (Figure 1.7). Farming and other practices gradually depleted non-reserves land until only approximately 5% of the high forest was outside the reserves (Figure 1.8). Currently, only approximately 20% of the high forest vegetation type remains. Table 1.3 lists the forest reserves in Ghana and summarizes their general characteristics.

Forest Plantations

Because of the steady decline of the forests outside the reserves, the Ghana Forest Services Division has established plantations in the forest as well as savannah areas across the country. The total forest plantation estate in Ghana, including private plantations, is currently estimated at 160,000 ha (FAO 2005). A national reforestation program launched at the beginning of 2000 had the goal to reforest 20,000 ha degraded forest reserves annually. There are four on-going forest plantation development initiatives namely: a) Community-based and poverty alleviation (Modified Taungya System) on forest reserves, b) Small farmer Agroforestry Scheme off of forest reserves, c) Employment focused plantations and d) Private sector led commercial plantation development. An estimated total of 107,000 ha have been established so far under these plantation projects.

Probably the most successful are those of teak (*T. grandis*) plantations (Plate 10). This species grows well in all regions of Ghana and produces excellent fuel-wood and termite-resistant poles. A large number of tree species have been introduced into Ghana with highly variable success (Appendix C). There is strong interest in the use of native species in plantations such as *Terminalia superba* (Plate 11) and the development of plantation strategies for mixed native species plantations (Plate 12). Insects have played a major role in limiting the establishment of several major species such as *Khaya* spp. and *Milicia* spp. in plantations. Nkansa-Kyere (1972) reported that 94 species of trees have been introduced into Ghana, many of them for plantation purposes. Common exotic species used in Ghana include *Azadirachita indica, Senna [Cassia] siamea, Eucalyptus* spp., *Gmelina arborea, Pinus* spp., *Tectona grandis,* and *Cedrela odorata* (Plate 13).

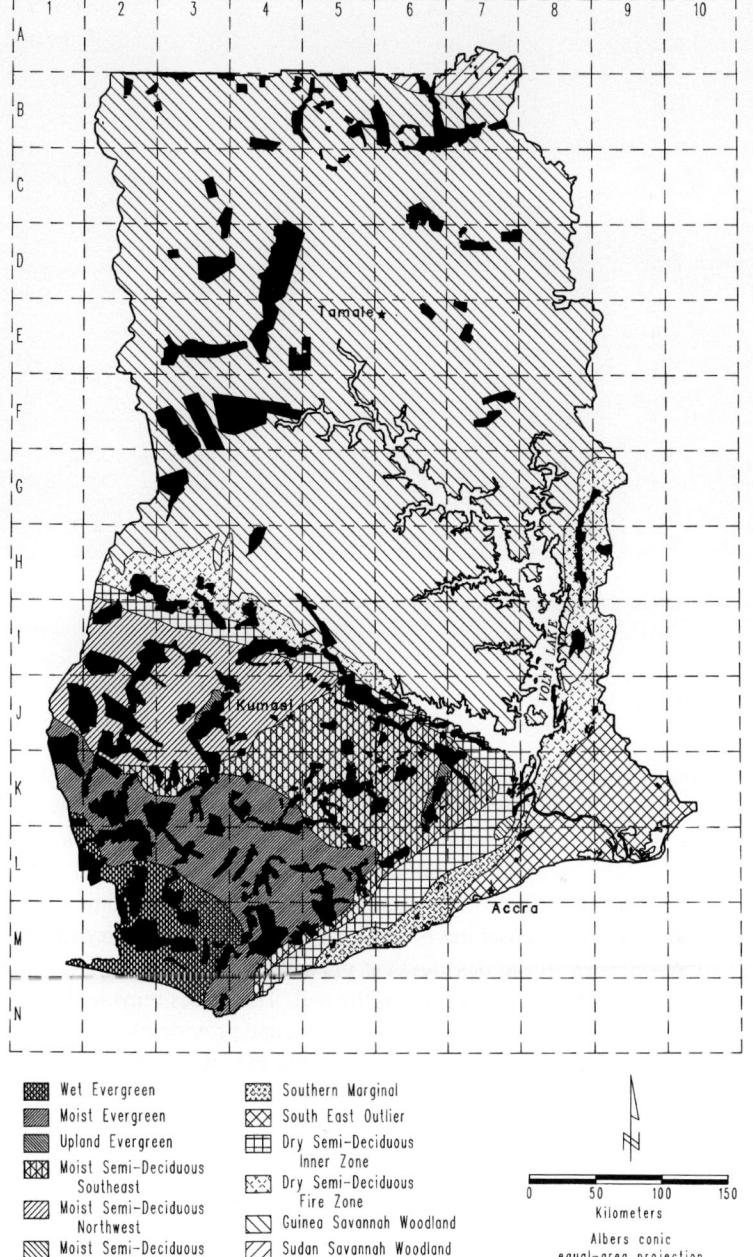

Figure 1.7 Areas within the closed forest and savannah that have been included in the Forest Reserve System in Ghana (*see color plate*)

Figure 1.8 Deforested landscape near Lake Bosumtwi in Central Ghana

Agroforestry

Agroforestry is generally considered as a specialized branch of forestry in which tree species are grown in association with agronomic crops. Others however, view agroforestry more broadly to encompass all practices that involve a close association of trees or shrubs with crops, animals, or grazing (Rocheleau et al. 1988). This association can occur concurrently (intercropping) or at different times in the same location (rotation). Alley cropping is a common type of agroforestry in which tree species (often nitrogen-fixing species) are interplanted with agronomic crops. The tree species bring nutrients from deep in the soil. Then trees recycle them through leaf fall, fix atmospheric nitrogen, reduce erosion, and provide fuel to cook the food crops. In Ghana this practice is relatively new and mostly experimental (Plate 14). Alley cropping has not been widely accepted by local farmers who prefer a more informal mixture of trees and agricultural crops. Taungya is an agroforestry system which combines the establishment of forest plantations with food crop production. Farmers plant and care for the tree species along with food crops. After a few years, when the tree crops are well established, farming is abandoned and the plantation is well established. The taungya system has been practiced in Ghana for many years. In areas where there is land scarcity for food crop cultivation, farmers are allowed access to degraded forest reserves for taungya farming. A modified taungya system that assured farmers of 40% of the proceeds from tree crops was introduced in 2005. Insects that attack and damage trees in an agroforestry area can be very serious pests.

Table 1.3 Forest reserves in Ghana

Reserve	Map index	Area (km²)	Forest type[a]
Abasuba	L6	0.47	SM
Abisu	J6	13.45	DSIZ
Aboben Hill	K8	6.77	SM
Aboma	I5	42.83	DSFZ
Aboniyere Shelterbelt	J2	49.56	MSNW
Abrimasu	I5	28.60	DSFZ
Abutia Hills	J8	8.61	
Achimotaabasuba	L7	1.66	
Afao Hills	K3	35.88	MSSE
Afia Shelterbelt	K5	20.55	MSSE
Afram Bukunaw	K7	11.33	
Afram Dawa	K7	8.87	
Afram Headwaters	I4	195.05	DSFZ
Afrensu-Brohuma	I4	75.26	DSFZ
Agali Stream Fuelwood	K9	10.95	
Ahirasu 1	L6	0.43	
Ahirasu 2	L6	0.90	SM
Aiyaola	K6	39.37	MSSE
Ajenjua Bepo	K5	4.46	MSSE
Ajuesu	L5	15.86	ME
Akrobong	L6	3.14	SM
Amama Shelterbelt	I3	48.10	MSNW
Ambalara	C3	145.57	
Anhwaiso East	K3	121.93	MSSE
Anhwaiso North	K3	7.58	MSSE
Anhwaiso South	K3	19.29	
Ankaful Fuelwood	M5	2.39	SM
Ankasa River	M2	506.83	WE
Anum Su	J5	37.94	MSSE
Anum Su Southern Sector	J5	7.64	MSSE
Apamprama	K4	34.86	MSSE
Aparapi Shelterbelt	I3	22.84	MSNW
Apedwa	K6	6.02	MS-UE
Apepesu	H9	62.81	DSFZ
Asenanyo River	K3	240.36	MSNW
Asibelik	B5	21.22	
Asin Apihanim	L5	13.95	ME
Asin Atandaso	L5	160.89	ME
Asonari	J5	2.23	DSIZ
Asubima	I4	88.51	DSFZ
Asufu Shelterbelt East	I4	6.87	DSIZ
Asufu Shelterbelt West	I4	24.82	MS-DS
Asukese	I2	279.74	MSNW
Asuokoko River	H8	152.38	DSFZ
Atewa Range	K6	219.01	UE-MS

(continued)

Table 1.3 (continued)

Reserve	Map index	Area (km²)	Forest typeᵃ
Atewa Range Extension	L6	26.41	UE-MS
Auro River	K5	42.29	MSSE
Awura	I5	154.59	DSFZ
Ayum	J2	121.58	MSNW
Baku	L5	7.77	MSSE
Banba Watershed	G3	361.19	
Bandai Hills	J6	192.18	DSFZ
Bediako	K6	5.37	
Bemu	L5	40.57	ME
Ben East	L4	26.12	ME
Ben West	L4	49.08	ME
Bia Shelterbelt	I2	34.44	MSNW
Bia Tano	J2	203.90	MSNW
Bia Tawya	K2	76.50	ME
Bia Tributaries North	J1	367.94	MSNW
Bia Tributaries South	K1	361.49	ME-MS
Bimbong	L5	121.89	ME
Bina River	L3	93.24	
Birim	L5	8.77	
Birim	L5	45.68	MSSE
Birim Extension	L5	25.81	MSSE
Bobiri	J5	55.52	MSSE
Bodi	K2	237.21	ME
Boi Tano	M2	142.75	WE
Bomfoum	J5	294.66	DS
Bon River	L2	285.39	WE-ME
Bonkoni	J2	69.82	MSNW
Bonsa Ben	L4	160.39	ME
Bonsa River	M4	151.05	ME
Bonsam Bepo	J2	124.74	MSNW
Bopono	B5	54.13	
Bosomoa	H4	149.64	
Bosumkise	I3	152.81	MSNW
Bosumtwi Range	K5	75.93	MSSE
Boti Falls	K7	4.61	MSSE
Bowige Range	L3	107,27	ME
Brimso	M5	12.63	SM-DS
Buligu	E7	52.62	
Bumbugu 1	A7	1.98	
Bumbugu 2	B7	7.74	
Cape Three Points	N3	48.45	ME
Chai River	G8	105.17	DSFZ
Chana Hills	B5	19.20	
Chasi	B5	66.30	
Chira	B4	42.77	

Table 1.3 (continued)

Reserve	Map index	Area (km²)	Forest type[a]
Chiremoasi Bepo	K5	8.30	MSSE
Chirimfa	I5	101.71	DSFZ
Dadiaso	K1	61.78	ME-WE
Daka Headwaters	D7	148.45	
Damongo Scarp	E4	32.78	
Dampia Range	K4	88.54	MS-ME
Dechida	L7	7.43	
Dede	J7	60.95	DSFZ
Dencheredama	D4	1,112.34	
Denyau Shelterbelt	K4	10.39	ME
Desiri	J3	139.80	MSNW
Disue River	L2	24.69	WE-ME
Dombe	E4	1.47	
Dome River	J5	79.21	MSSE
Draw River	M3	219.41	WE
Dunkwa Town Plantation	L4	24.25	
Dunkwill	E7	89.29	
Ebi River Shelterbelt	M3	29.57	WE
Esen Epa	L6	47.27	MSSE
Esuboni	L6	28.78	MSSE
Esukawkaw	K6	131.99	MSSE
Fiakonya	L7	7.64	
Fum Headwaters	K5	79.21	MSSE
Fure	M3	145.26	WE
Fure Headwaters	L3	172.12	WE
Gambaga Scarp East	B7	183.09	
Gambaga Scarp West	B6	157.26	
Gia	C5	20.98	
Gianima	I4	20.74	DSIZ
Goa Shelterbelt	J3	24.97	MSNW
Inchaban Fuelwood Plantation	N4	1.89	
Jade Bepo	K6	6.44	MSSE
Jema Asemkrom	M2	116.60	
Jeni River	K4	26.52	MSNW
Jimira	J4	57.14	MSNW
Kabakaba Hills East	J8	9.09	DSFZ
Kabakaba Hills North	J8	2.91	DSFZ
Kabakaba Hills West	J8	10.03	DSFZ
Kabo River	H8	98.82	DSFZ
Kade Bepo	K6	13.74	MSSE
Kajease	K6	28.60	MSSE
Kakum	M5	202.93	ME
Kamba	B2	40.75	
Kandebeli	B5	21.93	
Kanjaga Pumbisi	C5	9.33	
Karaga	D7	24.35	

(continued)

Table 1.3 (continued)

Reserve	Map index	Area (km^2)	Forest type[a]
Karkar Plantation	B2	2.52	
Keni Keni	E3	618.91	
Klemu Headwaters	J9	17.89	DSFZ
Kokotintin Shelterbelt	K5	10.00	MSSE
Komeda Fuelwood	M5	2.68	
Kpanda Range Dayi block	J8	24.24	DSFZ
Kpandu Range West	J8	40.09	DSFZ
Krochua	L5	11.89	ME
Krokosua Hills	J2	480.34	MSNW
Kronwam	J5	5.61	DSIZ
Kulpaln Tributaries	C3	99.32	
Kulpawn	B3	46.72	
Kumasi Town	J4	0.98	
Kumawa Water Supply	J5	1.26	
Kumbo	F7	138.81	
Kunda	H8	1.18	MSSE
Kunsimua Bepo	K5	9.52	DS-MS
Kwamisa	I4	86.95	MSSE
Kwekaru	K6	11.42	MSSE
Kwesi Anyinima	K6	1.68	
Laboni	34	271.05	
Lambo	F7	124.57	
Lanka	F3	597.76	
Larabanga Konkori	D4	474.46	
Lawra	B2	1.12	MSSE
Mamang River	K5	55.13	ME-WE
Mamiri	L3	60.57	DSFZ
Mankrang	I3	97.09	MSNW
Manzan	JI	74.19	
Mawbia	B4	140.97	ME
Minta	L4	28.35	MSSE
Mirasa Hills	K5	60.94	
Morago River East	B7	67.58	
Morago West	B7	41.84	MSNW
Mpameso	I2	338.92	MSNW
Muro	J2	61.75	
Nandom	B2	60.24	
Nasia Tributaries	C6	316.72	
Ndumeri	M3	69.08	WE-ME
Nueng	M3	137.49	
Nueng North	M3	29.00	
Ngoben Shelterbelt	L3	39.82	ME
Nkonto Ben	L4	16.62	ME-MS
Nkrabia	L4	100.61	MSSE
Nkwanda	K6	11.91	
No. Nome 3	B6	101.14	DSFZ
North Bandai Hills	J6	71.68	

Table 1.3 (continued)

Reserve	Map index	Area (km^2)	Forest type[a]
North Formangsu	J6	45.04	MSSE
Northern Scarp East	J6	46.01	DS
Northern Scarp West	J6	64.89	DS
Nsemre	H3	26.75	DSFZ
Nsuensa	K6	75.62	MSSE
Nuale	D3	43.68	
Numia	K5	52.49	MSSE
Nyamibe Bepo	K5	25.76	MSSE
Nyenbong 1	F4	3.17	
Obotumfo	L6	3.25	SM
Oboyow	L6	62.96	MSSE
Ochi Headwaters 1	L5	1.23	ME
Ochi Headwaters 2	L5	14.54	ME
Oda River	K4	163.85	MS-ME
Odome River	I8	14.44	DSFZ
Ofin Headwaters	I5	11.41	DSIZ
Ofin Shelterbelt	J3	54.83	MSNW
Ongwam 1	J5	21.25	DSIZ
Ongwam 2	J5	13.19	DSIZ
Ongwam 3	J5	2.17	DSIZ
Onuem Bepo	K5	33.13	MSSE
Onuem Nyamibe Shelterbelt	K5	28.91	MSSE
Onyimsu	J5	11.04	MSSE
Opimbo 1	L6	1.75	DSIZ
Opimbo 2	L6	0.38	DSIZ
Opon Mansi	L4	116.66	ME
Opro River	I4	129.30	DS
Owabi Water Works	J4	14.75	
Pamu Berekum	I2	188.28	DS-MS
Pkandu Plantation	J8	1.52	
Pogi	C5	27.79	
Poli	B3	32.09	
Ponro Headwaters	K4	13.29	MSSE
Pra Anum	K5	137.23	MSSE
Pra Birim	L5	12.42	MSSE
Pra Suhien 1	M4	73.03	ME
Pra Suhien 2	M5	106.22	ME
Prakaw	K5	10.42	MSSE
Proposed Greenbelt	K4	10.55	
Pru Shelterbelt	I5	59.35	
Pudo Hills	B4	46.63	
Red and White Volta West	B65	543.10	
Santomang	K2	8.55	
Sapawsu	K8	16.35	SM
Sawsaw	H3	75.46	DSFZ
Sekondi Waterworks 2	M4	5.80	MSSE
Sekondi Waterworks 3	M4	1.83	MSSE

(continued)

Table 1.3 (continued)

Reserve	Map index	Area (km²)	Forest type[a]
Sephe	D3	391.89	
Sinsablegwini	E6	69.43	
Sissili Central	B4	153.17	
Sissili North	B5	54.70	MSSE
South Chipa Tributary	L8	34.75	MSSE
South Fomangsy	J6	38.12	MSNW
Southern Scarp	J6	249.25	ME
Subin	J2	211.82	ME
Subin Shelterbelt	K4	26.86	ME-MS
Subri River	M4	598.25	ME-MS
Suhuma	K3	366.56	ME
Sui River	K2	311.36	MSSE
Sukusuku	J1	122.40	ME
Supong	L5	44.37	DSFZ
Suruma Shelterbelt	K4	24.68	DSFZ
Tain Tributaries 1	I3	26.78	
Tain Tributaries 2	H2	459.08	
Tamale Fuelwood 1	E6	0.34	
Tamale Fuelwood 2	D6	0.57	
Tamale Waterworks	E6	0.26	
Tanja	D7	115.78	
Tankara 1	B5	10.24	
Tankara 2	B5	3.45	ME
Tankwindo East	B5	191.66	ME
Tano Anwia	L2	152.98	WE-ME
Tano Ehuro	L2	212.56	MS-UE
Tano Nimiri	L2	233.22	MS
Tano Ofin	J3	405.70	MSSE
Tano Suhien	K3	81.04	MS
Tano Suraw	K3	25.01	
Tano Suraw Extension	K3	83.24	MSNW
Tapania	B3	100.82	DSFZ
Tinte Bipo	J4	125.84	ME
Togo Plateau	I8	144.38	ME
Tonton	K3	154.29	
Totua Shelterbelt	L3	83.23	
Trans Bia	J2	148.92	
Tumu	B4	44.51	
Upper Tamne Blocks 1 and 2	A7	4.66	
Upper Tamne Blocks 3	A7	10.19	
Upper Tamne Blocks 4	A7	1.42	
Upper Tamne Blocks 5	B7	3.24	
Upper Wassaw	K3	47.43	ME
Volta River	K7	47.12	DS-SM
Wassaw	K3	37.74	
Wawahi	L5	39.62	MS-ME

Table 1.3 (continued)

Reserve	Map index	Area (km²)	Forest typeᵃ
Wiaga	B5	113.26	
Wiaga	B5	12.21	
Winneba State	M6	2.02	
Worobong	K7	171.81	
Worobong North	J7	47.91	DS
Yakombo	F4	1,138.16	
Yaya	I3	54.60	DSFZ
Yendi Forest Plantation	E7	1.29	
Yendi Town Plantation	E7	0.87	
Yenku	M6	8.47	SM
Yeraba	F3	543.46	
Yogaga	K7	8.85	
Yoyo	L2	231.31	ME
Zawse	A7	1.91	
Zawsi Hills	A7	3.73	

ᵃSeveral former reserves which were abandoned when they were flooded by the Volta Lake have been omitted from the list. Areas given are total areas enclosed by the reserve boundary; no allowance has been made for permitted farms within the reserve. Forest type codes are found in the caption to Plate 1.

Forest Resource Condition

Ghana's forest resources contribute significantly to the national economy, generating about 6% of gross domestic product (GDP) annually. The sector employs about 75,000 people, with an estimated 2 million indirect dependents. In many rural communities the forest is the major source of livelihood for provision of fuel wood and a variety of non-timber forest products (NTFPs). Snails (*Achatina achatina*) are an important NTFP in West Africa (Plate 15). However, Ghana's forest resources are declining at an alarming rate and this is mainly due to years of unsustainable forestry and agricultural practices. Annual rates of deforestation are among the highest in the world (Wagner and Cobbinah 1993). Forest degradation has also led to about 40% reduction in run off and quality deterioration of major rivers such as Deusu, Pra, and Tano (Cobbinah per com). It is now widely accepted that traditional forest products such as mushrooms, snails, bush meat, medicinal plants, and wild fruits are more difficult to find today than perhaps in the past few decades.

Inventory analysis from the mid-1980s projected depletion of many valuable timber species within a period of about 25 years (Alder 1989). Species listed included Iroko (*Milicia excelsa*), Afrormosia (*P. elata*), and several mahoganies (*Khaya* and *Entandrophragma* spp.). Current estimates put the total forest cover loss (including reserved and off-reserve areas), between 1990 and 2005 at about 26% (FAO 2005). Within the same period the total growing stock decreased by about the same margin. Species such as *Ceiba pentandra, C. zenkeri*, and *Antiaris toxicaria* are being heavily exploited. The annual rate of forest cover loss in Ghana is 1.7% or about 120,000 ha/year (FAO 2005). A recent World Bank / ISSER / FORIG study of estimated cost of forest depletion as 3.5% of Ghana GDP representing 60% of the degradation of renewable natural assests. Forest depletion might lower Ghana's GDP growth which recorded a respectable 6.2% in 2006 of remediation is not taken.

Chapter 2
Defoliating Insects

Introduction

Insect defoliators have long been recognized as important pests in forestry. In West Africa, the main species of insect defoliators belong to three orders; butterflies and moths (Lepidoptera), grasshoppers (Orthoptera), and leaf beetles (Coleoptera). Whereas both the immature and adult stages of Orthoptera and Coleoptera cause feeding damage, only the larvae (caterpillars) of Lepidoptera are leaf feeders. It is estimated that more than 60% of important insect defoliators in Ghana are caterpillars.

When defoliation is moderate and evenly distributed over the crown of a tree, the damage may not be easily observed. In many cases defoliation might reach 50% or more before the tree would appear abnormal. A single defoliation, though severe, rarely causes mortality, with exception of evergreen conifers such as *Pinus* spp. which are rarely grown in Ghana. Deciduous conifers and hardwoods with large amounts of stored food and in vigorous physiological condition can produce new leaves quickly and minimize the effect of leaf loss. Repeated defoliation can, however, cause significant growth loss and may even result in death of the tree. The magnitude of the damage is a function of the extent of defoliation, timing of the attack, tree species involved, and general vigor of the attacked tree.

Defoliation reduces tree growth through the reduction of total photosynthetic area. With lower food-producing ability, the first parts of the tree to die are the small roots and twigs. If defoliation is heavy and repeated, death of the entire tree may follow. Occasionally outbreaks of defoliators arise rather suddenly and with no apparent warning. For instance, the *Terminalia ivorensis* plantation at South Fumangso Forest Reserve was severely defoliated for the first time in May/June 1984 by the notodontid *Epicerura pulverulenta* Hampson. The only realistic way to be prepared to control defoliator epidemics is to anticipate them at least one season in advance. Regular observations of the status of potentially dangerous pests could give a warning of approaching danger before the insects inflict any serious injury.

Outbreaks of Defoliating Insects

It can be generally stated that extensive outbreaks of defoliating insects are uncommon in the high forests of Ghana. This is true because the forests have a high degree of species diversity and most insects have a narrow host range. A few species of tree, like *Triplochiton scleroxylon*, occur naturally in singe species clumps and are subjected to periodic outbreaks of defoliators. When, however, single species plantations are established, the probability of an outbreak of a defoliating insect increases substantially. Many of the current defoliating pests have become more serious because of the tendency to establish single species plantations.

Types of Defoliation

Insect defoliators fall into three main groups: leaf miners, skeletonizers, and whole leaf feeders. Leaf miners feed upon the succulent interior leaf tissues as they tunnel between the upper and lower epidermis of the leaf. Skeletonizers eat all the leaf except the vascular portions, thus skeletonzing the leaf. The majority of the leaf-feeding insects, however, are whole-leaf feeders, which eat all the leaf tissues. Some defoliators are miners during a part of their developmental period and skeletonizers at a later time. Others are skeletonizers during their early stages and whole-leaf feeders during later stages. The immature stages of some whole-leaf feeders, particularly butterflies and moths, may spin silk, which is used to twist or distort the leaf. The insects feed, rest, develop, and pupate within this protective web of silk and leaves. Defoliating insects possess a pair of mandibles or true jaws which are tooth-like and are adapted for cutting and crushing food. They are referred to as having mandibulate or chewing mouthparts. We shall now turn our attention to some specific insect defoliators known to occur in Ghana.

Lepidopterous Defoliators

Anaphe venata Butler (Lepidoptera: Notodontidae)

This moth causes extensive defoliation of *T. scleroxylon* (Wawa, Obehe), a valuable timber tree in the natural high forests of Ghana. The insect is also known to occur in Nigeria and Cameroun. In Nigeria the larvae of this insect are roasted in dry sand and eaten by many local tribes (Ashiru 1988a). Related species *Anaphe reticulata* and *A. panda* are eaten fresh, boiled, roasted or are dried, and used as a powder in Tanzania, Zambia, and Zaire (Schabel 2006). Repeated annual defoliations have been recorded in the Bobiri Forest Reserve, east of Kumasi, during the months of August and September (Figure 2.1).

Figure 2.1 Adult specimen of *Anaphe venata* (wingspan 4.5 cm.), an important defoliator of *Triplochiton scleroxylon*

Description and Life History

A brief life history of this notodontid was given in Taylor (1960). Eggs are laid on the leaves of the *T. scleroxylon* trees in May. In general most of the eggs are laid on the tips of branches in the lower canopy (Ashiru 1988b).The caterpillars emerge in approximately 3 weeks and feed on the leaves, often stripping the tree. The brown, hairy caterpillars mature in August and then descend from the tops of the host plant in spectacular long processions. Once on the ground, they form communal cocoons on the undersides of leaves of surrounding shrubs and low trees. These cocoons are light brown, tough, and papery in consistency and variable in shape. The larvae remain in the prepupal condition for 2–3 months. Eventually, they pupate in separate cocoons within the communal cocoon. Adults emerge in April and mating takes place almost immediately (Figure 2.2). The females then lay their eggs on tall *T. scleroxylon* trees and the cycle continues.

Damage

A. venata has not yet been observed attacking *T. scleroxylon* plantations in Ghana. This may relate to the low abundance of wawa plantations. All *A. venata* attacks in the natural forests have been observed on tall trees that are 50 years or older. Thus, although the attack may be severe in some localities, the loss in increment is not apparent, and no effort has yet been made in Ghana to control the insect. Kudler (1967a, b), however, postulated that severe defoliation might affect seed production by the tree and future plantations may be impacted.

Figure 2.2   Adult *Godasa sidae* (wingspan 4.2 cm); the larvae of this insect skeletonize leaves of *Mansonia altissima*

Lamprosema lateritialis Hampson (Lepidoptera: Pyralidae)

Lamprosema lateritialis is the most serious pest of the valuable indigenous timber species Afrormosia, *Pericopsis elata* (Harms.) van Meeuwen. Although the genus *Lamprosema* Hubn. has a wide distribution throughout the tropics and several species have been reported as agricultural pests of leguminous plants in the Pacific Basin, *L. lateritialis* is the only species known to attack a forest tree. The insect is restricted to Central and West Africa, where it is widespread in the lowland rain forests of Nigeria, Ghana, and the Côte d'Ivoire. In Ghana, the moth was first observed in a nursery in Kumasi during 1965. The pest is widespread in the Afram Headwaters and the Bia Group of Forest Reserves, and is believed to be endemic wherever *P. elata* grows naturally.

Description and Life History

The adult moth is a small, yellowish-brown insect with a wingspan of approximately 18–21 mm (Figure 2.3). The forewings are yellowish with three wavy brown bands running more or less vertically across the width of the wings; the hind wings have two brown bands. The female lays spindle- or oval-shaped egg batches 2–4 mm wide and 5–12 mm long. Within this egg batch, individual eggs 0.8–1.2 mm long and 0.8 mm wide overlap one another (Plate 16). A cluster of viable eggs may contain between 30 and 200 eggs and may be laid on both the upper and lower surfaces of the afrormosia leaflet. At Mesewarm, over 80% of the eggs were laid on the upper surface of leaflet. When freshly laid, the egg batch is cream colored but turns brown

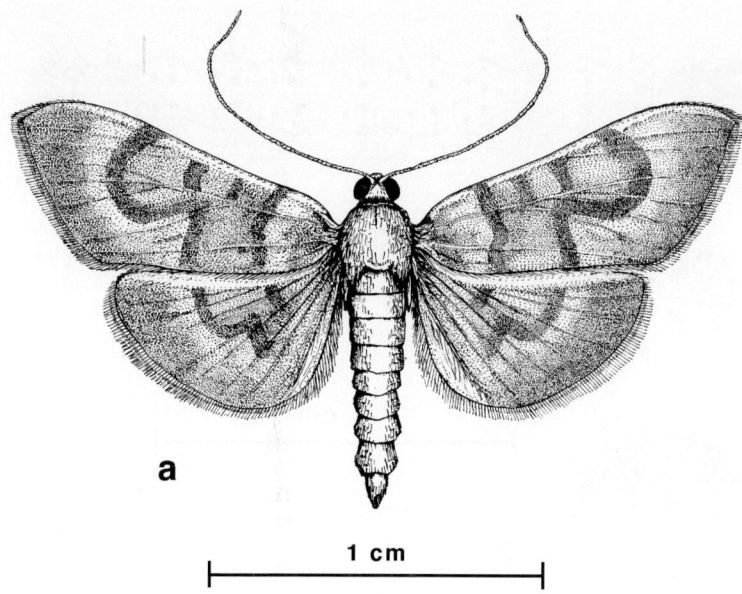

a

|——————— 1 cm ———————|

Figure 2.3 *Lamprosema lateritialis* adult. (Drawing courtesy of J. Kudler.)

after 24–48 h. Infertile eggs are pale green and do not undergo any color change. Each egg colony is protected by a thin, transparent, waxy substance, which hardens when exposed to the air. The incubation period is 7–9 days.

Once the eggs eclose, all the caterpillars from an egg batch move together, possibly by "communal stimulation," to search for a suitable nesting site which is normally provided by two or more overlapping leaflets. The searching distance is usually a few centimeters from where the eggs hatched. After reaching the target leaflets, the fragile-looking caterpillars (each approximately 1.9 mm long) settle on the adaxial (upper) surface of the substrate leaflet. A few insects then position themselves along the edges of the leaf blade and, by repeatedly touching the opposite edges of the adjacent leaflet with their mouthparts, the caterpillars are sandwiched between the two leaflets by means of silk. This simple two-leaf nest is typical of the early instars of *L. lateritialis* (Plate 17 and 18).

The insects now feed gregariously; skeletonizing the surfaces of the leaves. The damaged leaves turn brown and wither in about 5–6 days. The larvae then crawl out the damaged leaves to the nearest pair of suitable fresh, mature leaves where the process of nest building and feeding is repeated. As the caterpillars grow, crowding becomes a problem and the simple nest is no longer adequate. The fourth and fifth instar caterpillars (Figure 2.4, Plate 20) by virtue of their larger sizes (11–16 mm) and highly sclerotized mouthparts, are capable of pulling together more leaves to construct an elaborate sac-like, compound nest (Plate 18). Considerable damage to the plant occurs at this stage (11–16 days after egg eclosion). The caterpillars feed freely on the inner core of leaves, while the outer cover of leaves is neatly knitted

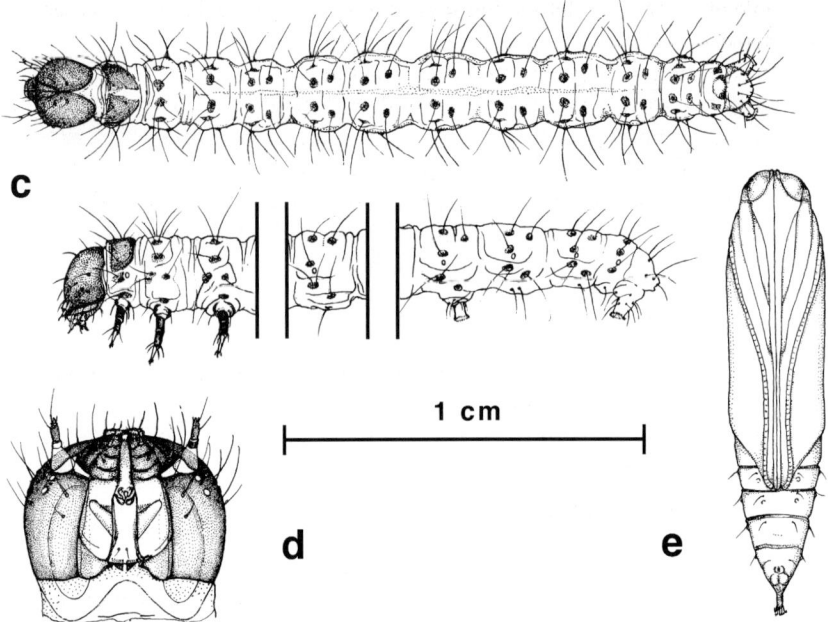

Figure 2.4 *Lamprosema lateritialis* caterpillar (dorsal and lateral view), head, and male pupa. (Drawing courtesy of J. Kudler.)

together in an overlapping fashion to render the nest waterproof and almost impervious to intrusion by predators. As feeding continues, the nest withers. *Lamprosema lateritialis* normally pupates in the withered, rolled-up nests which are held in position by dense strands of silk.

In the laboratory, the developmental periods of the first, second, and third larval instar stages are 5–7, 3–5, and 3–4 days, respectively, while the developmental stages of the fourth and fifth instar stages are 3–4 and 5–7 days, respectively. The prepupal stage lasts 24 h and the pupal stage lasts 6–8 days. Time from egg to adult is 34–45 days; 8 or 9 generations are produced in 1 year.

In the field at the FORIG plantations at Mesewam, insect numbers closely follow the annual rainfall pattern. Egg numbers are low in the dry months of December/February, but insect populations begin to build with the onset of the rains in March, reaching a peak in May/June. Populations decline in August, but rise again in early September to reach a second (minor) peak in October.

Damage

Within the average larval life span of 21 days, approximately 140 mg dry weight of leaves is consumed by each caterpillar. This is equivalent to 2–3 pinnate leaves. At this rate, caterpillars emerging from one healthy, normal-sized egg batch would be capable of completely defoliating several 6- to 7-month-old afrormosia seedlings in the nursery (Plate 19).

At Mesewam, seedling loss due to repeated defoliation of afrormosia by caterpillars of *L. lateritialis* was estimated at 30–40% in a period of 1 year. Diameter growth among saplings was very sensitive to foliage loss and when the plants were subjected to a relatively low 25% defoliation at 6 weekly intervals for a period of 1 year, there was a marked reduction in girth growth of approximately 35%. The loss in height growth for the same period was a meager 8% (Atuahene 1989).

Pest Management

Very few controlled field studies have been conducted on *L. lateritialis*. Several larval and pupal parasites have been collected in Ghana (Kudler 1967a, b; Atuahene 1983), including *Neoplectops nudinerva* Mesnil, *Pseudoperichaeta* sp., *Palexorista* sp. (Diptera: Tachinidae), *Tetrastichus* sp. (Hymenoptera: Eulophidae), *Brachymeria paolii* Masi, *Braunsia erythraea* (Hymenoptera: Chalcididae), *Braunsia* sp. nr. *analis* Kriebch, *Cardiochiles* sp., *Meteoride hustsoni* Nixon (Hymenoptera: Braconidae), *Xanthopimpla* sp. (Ichneumonidae), and an unidentified red mite (Ectoparasite). The entomogenous fungus *Beauveria bassiana* (Bals.) Vuill. (Fungui Imperfecti) was found attacking both caterpillars and pupae of *L. lateritialis*. Even though the list of parasites looks impressive, the level of parasitization in the field has been consistently low, seldom exceeding 20% in any month of the year.

Microbiological control using the entomophagous bacteria *Bacillus thuringiensis*, Berliner has shown good results at Afram Headwaters Forest Reserve. Suspensions of *B. thuringiensis* dispersed by knapsack sprayers at the rates of 5×10 and 5×10 spores per acre in 22 gallons of water were effective. Laboratory tests indicate that caterpillars of *L. lateritialis* are very susceptible to the fungus *Beauveria*, and the LD_{50} (Lethal dose to kill 50% of the population) value for fourth instar larva is estimated at 2.3×10 conidiospores per insect.

In the nursery, afrormosia seedlings are vulnerable to attack by *L. lateritialis*. Since over 80% of *L. lateritialis* eggs are laid on the upper surfaces of *P. elata* leaves and the eggs may be easily recognized by any unskilled laborer, it is recommended that eggs be removed from seedings during normal morning waterings. Using this mechanical method of control, mortality of nursery seedings due to *L. lateritialis* larvae could be reduced from 30–40% to 5% or less in a year.

Strepsicrathes rhothia Meyrick (Lepidoptera: Tortricidae)

Strepsicrathes rhothia is a major pest of *Eucalyptus* spp. grown in Ghana. This pest was first detected during the early 1970s at Yenku Forest Reserve in the Coastal Savannah vegetation. Damage to the *Eucalyptus* nursery at Yenku has exceeded 50% in some beds, particularly towards the beginning of the dry season. The insect also occurs at Mesewam, near Kumasi, in the moist, semi-deciduous vegetation type.

Description and Life History

The larva rolls a single eucalyptus leaf, forming a shelter in which it feeds and rests. When the leaf turns brown and withers as a result of the feeding, the larva crawls out to a fresh leaf to repeat the process. When the larva is fully developed, it pupates in the rolled-up leaf. In the laboratory, the entire life cycle requires approximately one month to complete. The egg stage lasts 3–4 days, the larva develops in 2–3 weeks and the pupal stage lasts 5–8 days. Adults are short lived, surviving only 2–3 days on sugar solutions in the laboratory.

Damage

All the four species of *Eucalyptus* introduced at Yenku Forest Reserve and at the FORIG plantations, Mesewam, are susceptible to *S. rhothia* attack. *E. tereticorinis* is the most preferred host species, with *E. alba, E. cadambae*, and *E. citriodora* preferred in descending order. The rate of infestation was about twofold more on shaded beds than on unshaded beds.

Pest Management

Both organochlorine and organophosphous insecticides have been used to control *S. rhothia*. When the insect population was high at Yenku, Bidrin, an organophosphorous compound, was particularly effective when applied at low concentrations. Aldrex 40 at 0.5% was also effective.

Godasa sidae Fabricius (Lepidoptera: Arctiidae)

The insect has been observed skeletonizing leaves of *Mansonia altissima* in the nurseries at Opro, Afram Headwaters, and Afrensu Brohuma Forest Reserves and at the FORIG nursery, Mesewam, near Kumasi (Figure 2.2). The larvae are also responsible for the occasional heavy defoliation of the 60 acre *M. altissima* plantation at Jimira Forest Reserve.

Description and Life History

Eggs are laid in clusters on the underside of leaves. Foahom (1994) recorded between 56 and 250 eggs/cluster in Cameroon. Mature larvae are 4–5 cm long, pale yellow in color with black heads, and have black segmental spines and body markings on the dorsum. Six larval instars have been recorded. Larvae leave the trees to pupate in the litter on the forest floor. Development time from eclosion to pupation is 22–28 days, but may be more or less depending on locality and season (Foahom and Du Merle

1993). Pupae are naked and pale yellow in color with some black markings. The pupal stage generally lasts 10–13 days; however, Roberts (1969) reports that individual insects pupating in the early dry season may remain in the pupal stage for up to 32 days. There are between 7 and 10 overlapping generations per year.

Epicerura pulverulenta Hampson (Lepidoptera: Notodontidae)

This notodontid moth was responsible for a series of defoliation on young plantations of *T. ivorensis* grown in the South Fumangso Forest Reserve and on isolated plants found on the university campus in Kumasi during the rainy season of 1984. The major attacks occurred in May/June and a second, minor attack was observed in August/September. Initially, the only visible signs of damage were patches of defoliation at the tops of the affected trees and tiny pellets of excreta scattered beneath the tree. As the caterpillars matured, the pellets became larger and the defoliation spread to all crown levels of the trees. After 3 weeks the attacked trees were completely defoliated. In Nigeria, *E. pulverulenta* is reported to defoliate *T. ivorensis, T. superba, T. catappa* and *Annogeissus leiocarpus* (Akanbi 1990). In May 2004, *E. pulverulenta* attack occurred on *T. superba* seedlings at the Forestry Research Institute of Ghana nursery at Fumesua resulting in over 70% defoliation (Bosu et al. 2004). In Côte d'Ivoire, the *Epicerula* species that defoliates *T. ivorensis* and *T. superba* has been identified as *E. pergrisea* Hampson.

Description and Life History

Eggs of *E. pulverulenta* have not been observed on *T. ivorensis* trees. According to Akanbi (1990) the egg "cushion" is oblong, rarely, roundest, brown, and woolly and measuring about 1.02 cm and 0.75 cm at the widest area. Individual eggs are round to oval with an average width of approximately 0.52 mm. When freshly laid the chorion is ash, but turns brown at maturity.

Caterpillars are approximately 25–30 mm long and 1.5–2.0 mm wide, and when fully grown are light brown in color with dark stripes from the dorsal aspect of the longitudinal thorax to the last abdominal segment. Mature caterpillars descend to the ground to pupate; heavy mortality due to predation by birds and lizards occurs at this stage. Laboratory experiments indicate that a high proportion (approximately 80%) of pupae fail to emerge for unknown reasons. Three to four overlapping generations have been reported (Akanbi 1990).

Pest Management

Epicerula pulverulenta suffers very high mortality in natural populations due to the activities of natural enemies. In one evaluation, egg parasitism by *Gryon* and *Telenomus* spp. (Hymenoptera: Scelionidae) accounted for 33% mortality. Larval parasitism by a complex of parasitoids including *Carcelia* (*Senometopia*) sp. and *Zygobothria atropivora* was 4%. Other important parasitoids, larval parasitoids, are *Palexorisia* sp. *P. quadrizomula, Carcelia* sp., *Exorista xanthospsis* and *Perliampus dubius*. The entomogenous fungus *Paecilomyces farinosus* and the nematode *Hexamermis* sp. killed 27% and 2% of larvae, respectively. Predation by birds, lizards, spiders, and possibly viruses has been observed (Akanbi 1986; Mantanmi 1988).

In Côte d'Ivoire insecticidal trials conducted on *E. pergrisea* in the laboratory gave satisfactory results. A 9.6 g active ingredient per hectare concentration of decathrin (Decis) and 300 g active ingredient per hectare of hydrogenoxalat of thiocyclam (Evisect S) resulted in 99% and 94% mortality, respectively (Kanga and Fediere 1991).

Cirina forda Westwood (Lepidoptera: Saturniidae)

Cirina forda is the most serious defoliator of sheanut trees *Vitellaria paradoxa* that grow widely in the Guinea savannah belt of West Africa. The fruit kernel is an excellent source of edible fats, popularly known as shea butter which has a variety of domestic and industrial uses. (Plate 23). Heavy annual defoliation of up to 100% frequently occurs in Ghana and Nigeria (Dwomoh 2003; Odebiyi et al. 2003) (Plate 21, 22).

Description and Life History

Ande and Fasoranti (1997) have studied the life history of *C. forda* under laboratory conditions. Eggs are laid in clusters which hatch within a period of 30–34 days. The larvae feed gregariously and go through 5–6 instars over a period of 42–50 days. The final instar larva can attain up to 7.8 cm in body size and 0.72 cm head capsule width prior to pupation (Odebiyi et al. 2003). Mature larvae descend to the base of the tree to pupate in the soil. Fully developed pupae remain in diapause until they emerge the following wet season. Thus, *C. forda* produces only one generation per year (univoltine).

Natural control agents reared from field collected larvae in the laboratory include the parasitoids *Hockeria crassa, Megaselia scalaris*, and the parasitic worm *Gordius aquaticus* (Dwomoh et al. 2004). The fungi pathogens *Fusarium solani, Trichoderma* sp., *Aspergillus niger*, and *A. flavus* have been isolated from pupating larvae. However, the role of natural enemies in the population dynamics of *C. forda* is not known. The larvae of *C. forda* are important food source among some local people in Nigeria (Ande and Fasoranti 1997).

Miscellaneous Defoliators

In the previous section we have discussed some of the better known forest insect defoliators in Ghana. Some additional species we recognize as important include: *Maurilia arcuata* (Figure 2.5), *Streblote vesta* (Figure 2.6), *Anua subdivisa* (Figure 2.7), and *Phryneta leprosa* (Figure 2.8).

Figure 2.5 *Maurilia arcuata* (Lepidoptera: Noctuidae), a defoliator of various trees in Ghana

Figure 2.6 *Streblote vesta* (wingspan 7 cm) (Lepidoptera: Lasiocampidae), a defoliator of *Cupresus macrocarpa* on the campus of the University of Science and Technology, Kumasi

Orthopteroid Leaf Feeder

Zonocerus variegatus Linnaeus (Orthoptera: Acrididae)

This variegated grasshopper has recently emerged as a very serious pest of agricultural and agroforestry crops in Ashanti and Brong Ahafo regions of Ghana (Plate 25). The insects are particularly active during the dry months of November/February. Since 1984, there have been regular outbreaks of the pest every year in various localities resulting in the destruction of crops.

Figure 2.7 *Anua subdivisa*, a defoliator of several tree species (wingspan 5.8 cm)

Figure 2.8 *Phryneta leprosa* adult (3.5 cm. in length). These beetles feed on *Milicia excelsa* and *M. regia* in nursery and transplant beds

Description and Life History

The eggs are laid in the soil under undisturbed vegetation. Each female is capable of producing 1–4 egg pods, each containing 20–90 eggs. The eggs hatch in 6–7 months, depending on the environmental conditions, and the young nymphs feed gregariously by day and roost in shrubs at night. These grasshoppers tend to migrate slowly by walking rather than hopping. The adult grasshopper is rather sluggish and less gregarious than the nymphs and, therefore, less injurious. Their migration is influenced by population density, vegetation type, and micro-climate conditions.

Damage

During migration, the insects may devour virtually any green vegetation. On food farms, cassava (*Manihot esculenta* Crantz) leaves and bark are particularly

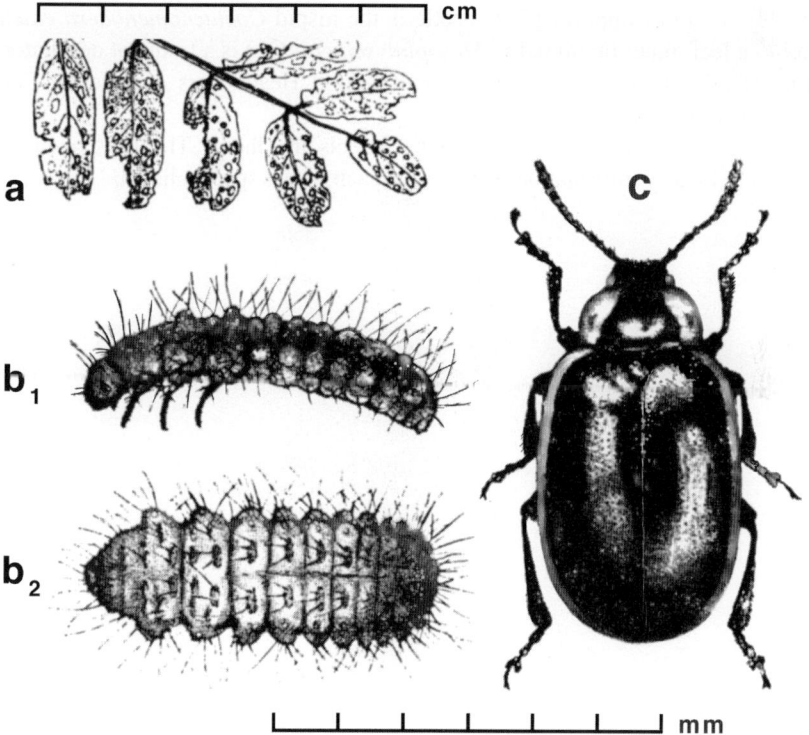

Figure 2.9 *Mesoplatys cincta* (Coleoptera: Chrysomelidae), an important defoliator of *Sesbania grandiflora* leaves: (a)defoliation; (b1)larva (lateral view); (b2) larva (dorsal view); (c)adult. (From Kudler 1970a–c; reprinted with permission of FPRI.)

favored, and may be completely stripped. One forest crop which is frequently defoliated is *Tectona grandis*, which is grown in the drier parts of the high forests (Plate 24). In 1987, the agroforestry plantation on the university campus in Kumasi was severely attacked by *Z. variegatus*. Both the cassava and the mulch crop *Gliricidia* spp. (Papillionaceae) were severely damaged.

Pest Management

Sporadic applications of various insecticides by individual farmers have been used to control *Z. variegatus*. However, no proper pest-management procedures have been implemented for this insect.

Leaf-Feeding Beetles

Few serious leaf beetles have been recorded on the forest tree species in Ghana. Probably the most important leaf beetle is the hispid *Coelaenomenodera elaeidis* Maulik, a leaf miner on oil palm. *Mesoplatyscincta* (Ol.) is a frequent defoliator of young stands of *Sesbania grandiflora* in Afram Headwaters and Yenku Forest Reserve (Figure 2.9).

Table 2.1 lists the known defoliating insects in Ghana. The host plants and known life history information is given for each insect species listed.

Table 2.1 Defoliating insects of Ghanaian forests

Species	Order: family	Host plant	Life history/comments
Achaea catella Guenee	Lepidoptera: Noctuidae	*Afzelia africana*	Larvae are defoliators, only light defoliation recorded
Achaea lienardi Boisduval	Lepidoptera: Noctuidae	Forsyth (1966) lists more than 40 host plants	Larvae are defoliators; adults are fruit piercers
Acraea pharsalus Ward.	Lepidoptera: Nymphalidae	Cacao	Defoliator
Anaphe venata Btl.	Lepidoptera: Notodontidae	*Triplochiton scleroxylon*	
Anomis leona Schaus.	Lepidoptera: Noctuidae	*Ceiba pentandra, Nesogordonia papaverifera, Sterculia rhinopetala, T. scleroxylon,* cacao	Minor defoliation recorded; larvae are polyphagous; collected in nurseries in Nigeria, also in Ghana
Anua ophiusa (Cramer)	Lepidoptera: Noctuidae	*Alchornea cordifolia, Combretum paniculatum, Phyllanthus discoideus*	(Forsyth 1966)
Ascotis reciprocaria (Walker)	Lepidoptera: Geometridae	*C. pentandra*	Larvae are defoliators
Ascotis selenaria reciprocaria (Walker)	Lepidoptera: Geometridae	Cacao and *C. pentandra* in Ghana.	Found on annual agricultural plants in Ghana (Roberts 1968; Smith 1965)
		On various hosts, especially *Eucalyptus* spp. and *Tectona grandis* in Nigeria	
Colocleora divisaria (Walker)	Lepidoptera: Geometridae	Various exotics in Nigeria, cacao in Ghana	(Forsyth 1966; Smith 1965)
Coelaenomenodera elaeidis (Maulik)	Coleoptera: Hispidae	*Elaeis guinensis*	Leaf miner, very serious
Coeliades forestan (Stoll)	Lepidoptera: Hesperiidae	Cacao, *Malpighia glabra*	Larvae are defoliators (Forsyth 1966; Smith 1965)
Dasychira georgiana Fawc.	Lepidoptera: Lymantriidae	*Terminalia superba*	Feeds on young leaves of host plants in plantations and in Mesewam nursery
Dasychira rhabdota Collenette	Lepidoptera: Lymantriidae	*Cola gigantea* in Nigeria *Cola acuminate* in Ghana	Defoliators of large trees (Eidt 1963; Forsyth 1966)

(continued)

Table 2.1 (continued)

Species	Order: family	Host plant	Life history/comments
Deilephila nerii Linneaus	Lepidoptera: Sphingidae	*Mitragyna stipulosa, Rauwolfia* sp.	Polyphagous on foliage of dicotyledons
Diacrisia attrayi Roths.	Lepidoptera: Arctiidae	*Cacao, Albizia zygia*	Larvae are defoliators
Diacrisia aurantiaca Holland	Lepidoptera: Arctiidae	Cacao	Larvae are defoliators
Diacrisia investigatorum	Lepidoptera: Arctiidae	*T. grandis*	Defoliator at Mesewam nursery
Diacrisia lutescens (Walker)	Lepidoptera: Arctiidae	*Cassia siamea, Dalbergia sissoo, Eucalyptus camaldulensis, E. deglupta, E. rudis, E. torelliana, Gmelina arborea*	Larvae common in nurseries and plantations in Nigeria on 1- and 2-year-old seedlings (Roberts 1969)
Earias biplaga (Walker)	Lepidoptera: Noctuidae	*Cacao, Sterculia tragacantha*	Unshaded seedlings
Epanaphe molonyi (Druce)	Lepidoptera: Notodontidae	*Byrsocarpus coccineus Cassia* sp., *T. scleroxylon*	(Forsyth 1966) Larvae are polyphagous defoliators (Browne 1968)
Epicerura pulverulenta Hamps	Lepidoptera: Notodontidae	*Terminalia ivorensis*	Defoliators; South Fumangso, UST campus 1984
Euproctis fasciata (Walker)	Lepidoptera: Lymantriidae	Variety of exotics, especially *Eucalyptus* spp. cacao in Ghana *Araucaria* and *Acacia* in Uganda	(Browne 1968)
Godasa sidae Fabricius	Lepidoptera: Arctiidae	*Mansonia altissima, Cedrela odorata*	Leaf skeletonizer of host plants, Afram Headwaters Forest Reserve Jemira Forest Reserve, Mesewam Polyphagous larvae. Occurs during November–March in Nigeria
Imbrasia nictitans (Fabricius)	Lepidoptera: Saturniidae	*Funtumia elastica, Holarrhena floribunda* Occasionally cacao	
Lamprosema indicata (Fabricius)	Lepidoptera: Pyralidae	*Glycine max*	Larvae webs foliage of soybean in Kumasi
Lamprosema lateritialis	Lepidoptera: Pyralidae	*Pericopsis elata*	Defoliator, builds nests with leaves of host plant Parasites: *Neoplectops nudinerva* Mesnil. *Pseudoperichaeta* sp. (Diptera: Tachnidae), *Trichogramma* sp. (Trichogrammiatidae), *Tetrastichus* sp. (Eulophidae), *Brachymeria* sp.

Species	Order: Family	Host plant	Remarks
Lechriolepis spp.	Lepidoptera: Lasiocampidae	*T. ivorensis*	(Chalcididae), *Braunsia* sp. *Analis* Kriebch, *Cardiochiles* sp. (Braconidae), *Palexorista* sp. (Tachinidae), *Brachymeria paolii* Masi, *Brachymeria erythraea* Masi (Chalcididae), *Meteoridea hutsoni* Nixon, *Xanthopimpla* sp. (Ichneumonidae), an unidentified red mite (Ectoparasite), *Beauveria bassiana* (Bals.) Vuill. (Moniliaceae: Fungi Imperfecti)
Lophocrama phoenicochlora Hamps	Lepidoptera: Noctuidae	*T. scleroxylon*	Defoliator
Maurilia phaea Hamps.	Lepidoptera: Noctuidae	*T. ivorensis*	Defoliator at Mesewan nursery
Megaleruca griseosericans	Coleoptera: Galerucidae	*Piper guineense*	Defoliator
Mesoplatys cincta (Ol.)	Coleoptera: Chrysomelidae	*Sesbania grandiflora*	Larvae feed on host plant (Forsyth 1966)
Nadasai splendens Druce	Lepidoptera: Lasiocampidae	*Entandrophragma angolense* in Nigeria *A. cordifolia* in Ghana	Defoliator of young trees in coastal shrubland
Nudaurelia dione (Fabricius)	Lepidoptera: Saturniidae	Cacao	In Ghana In Nigeria
Orgyia basali affinis (Holland)	Lepidoptera: Lymantriidae	*T. superba*, *Cordia milleni* in Ghana and Nigeria. Various exotics (*Casuarina equisetifolia*, *E. camaldulensis*, *E. deglupta*, *T. grandis*)	
Palpita ocellata Hamps	Lepidoptera: Pyralidae	*Funtumia ellastica H. floribunda*	Leaf roller; Mesewam nursery
Papilio demoleus Linnaeus	Lepidoptera: Papilionidae	Citrus	Occasional defoliator (Forsyth 1966)
Parasa viridissima Holland	Lepidoptera: Limacodidae	Cacao	
Parastichtia sp.	Lepidoptera: Noctuidae	*Cassia siamea*	Defoliating plantations of *C. siamea* Obuasi. Parasites: *Mesochorus* spp. (Ichneumonidae). *Proctomicroplitis fasciipennis* (Gahan) (Braconidae) Predators: *Demarius parvue* Dist. (Pentatomidae), *Rhinocoris* spp. (Reduviidae)

(continued)

Table 2.1 (continued)

Species	Order: family	Host plant	Life history/comments
Phryneta leprosa (Fabricius)	Coleoptera: Lamiidae	*Milicia excelsa, Milicia regia*	Newly emerged adults browse new shoots and foliage of young *C. excelsa* and *C. regia* in nurseries and transplant beds; also strip bark from main stems and lateral branches, damage found in the rain forests of Nigeria and East Africa (Roberts 1969; Gardner 1957)
Phymateus kazschi Bol	Orthoptera: Acrididae	*Eucalyptus* spp. *E. citriodora, E. robusta E. tereticornis*	General defoliators in Northern Ghana (Forsyth 1966) In Nigeria
Pimelephilia ghesquieri Tams	Lepidoptera: Pyralidae	*Elaeis guineensis*	Larvae feed on fronds of host plant at Northern Scarp, Mpraeso, pupal parasite: *Syntomosphyrum phaeosoma* Wtrst. (Hymenoptera: Eulophidae)
Schalidomitra remota Druce	Lepidoptera: Noctuidae	*Ficus capensis, Bombax*	(Forsyth 1966)
Serica sp.	Coleoptera: Melolonthidae	*T. scleroxylon*	Beetles found feeding on the foliage of seedlings at Aburi
Streblote vesta Druce	Lepidoptera: Lasiocampidae	*Cupresus macrocarpa*	Defoliator at UST campus. Outbreak occurred in 1971/1972, but insects have since disappeared
Strepsicrates rhothia Meyr.	Lepidoptera: Tortricidae	*Eucalyptus* spp.	Leaf roller and defoliator of *Eucalyptus* spp. seedlings at Yenku and Mesewam
Syagrus sp.	Coleoptera: Eumolpidae	*Leucaena glauca*	Beetles found defoliating *L. glauca* at Kpong
Trichotaphs sp.	Lepidoptera: Gelchiidae	*T. ivorensis*	Defoliator
Zonocerus variegatus Linnaeus	Orthoptera: Acrididae		Polyphagous feeder of considerable economic importance

Chapter 3
Sap-Feeding Insects

Introduction

The insects that we discussed in the previous chapter feed upon the foliage of trees by ingesting the solid parts. There is another important group of phytophagous insects which live upon plant sap. These sap-feeders have haustellate or sucking mouthparts in which the mandibles and maxillae have become slender bristle-like organs enclosed in a sheath formed by the labium. The mouthparts thus form a beak which is used to pierce the tissues and suck the fluid from the leaf.

Sap-feeding insects may injure plants directly by depriving the plants of its food and water supply. Indirect damage from feeding may cause gall formation, foliage disturbances such as bleaching or yellowing, and deformation such as leaf curling. Some sucking insects, such as aphids, may disseminate plant diseases. Generally, however, the effect of sucking insects upon trees is much less conspicuous than the effect of defoliators. The few exceptions will be treated in this chapter.

Sap-feeders belong to the order Hemiptera (formerly Hemiptera and Homoptera). Hemipterans seldom occur in sufficient numbers to be serious pests of forest trees. An exception is members of one family (Miridae), which are very serious pests on the forest crop cacao (*Theobroma cacao* Linnaeus).

The most important sap feeders in Ghana are members of the family Psyllidae being particularly harmful to some important indigenous timber species. Psyllids are small insects about the size of aphids. They are usually very active, moving rapidly in the combination of leaping and flying, but are incapable of sustained flight. The hind legs, which are used for leaping, are larger and more muscular than the other leg pairs. The venation in the wing is simple and there are a few marked deviations in wing venation among various genera.

Two genera of psyllids are particularly important to forestry in Ghana. These are *Phytolyma* spp. Scott, and *Diclidophlebia*.

Phytolyma spp. Scott (Hemiptera: Psyllidae)

This genus is found in tropical Africa throughout the range of *Milicia excelsa* and *M. regia*. Both plants species are extremely valuable in Ghana for their durable timber, and *Milicia* numbers are dwindling. Efforts by the Forest Services Division (FSD) of the Forestry Commission (FC) to establish plantations have been unsuccessful due primarily to attack by *Phytolyma* spp.

Previously, it was believed that a single species, *Phytolyma lata* Walker, was widely distributed and infested both *M. excelsa* and *M. regia*. But it is now evident that *P. lata* breeds successfully only on *M. regia* in Sierra Leone and possibly eastward to Ghana. From eastern Ghana to Tanzania, on the east coast of Africa, the predominant plants species is *M. excelsa* which is infested by another *Phytolyma* sp. or possibly by a complex of species.

In Ghana, where the two *Milicia* species overlaps, three *Phytolyma* species, *P. lata, P. tuberculata* (Alibert), and *P. fusca* Walker, have been collected at the Forestry Research Institute of Ghana nursery, Mesewam, near Kumasi. In areas dominated by *M. regia*, the most common of these insects' species is *P. lata* (Plate 26) and the least common species is *P. tuberculata*. On *M. excelsa*, the predominant species is *P. fusca*.

Description and Life History

Apart from their host relationships, the habits of the three species are very similar. The eggs are laid in rows or, more rarely scattered singly on buds, shoots, or leaves of the host plant. Each egg is approximately 0.27 mm long and 0.08 mm wide, creamy white in color, and has dark eyespot at one end. After approximately 8 days of incubation, the first instar nymphs emerge from the eggs and crawl over the plant surface. The young crawlers, as they are called, penetrate into the leaf tissues, breaking down the epidermal cells which cause fermentation of the leaf parenchyma. A gall is formed within 1 or 2 days which completely encloses the nymph (Figure 3.1, Plate 27). According to Browne (1968), galls developed on buds are more or less globular and smooth; some galls have remnants of the buds scales. Leaf galls are globular to ovate and most commonly occur on the midrib, while galls formed on stems or shoots are somewhat oblong. The insects feed within the gall tissue, which is many times larger than the insects within. Several such galls on young leaves and shoots may coalesce and become a mass of gall tissue (Plate 28). Five nymphal instars have been identified (Orr and Osei Nkrumah 1978) over the developmental period of approximately 2–3 weeks. The gall then becomes turgid and splits open (Plate 29), usually at the point of the original attack, to release the adults. Occasionally, however, the gall may harden without opening and the trapped insect dies. This usually happens when *P. lata* attempts to infests *M. excelsa* or when the other species of *Phytolyma* attacks *M. regia*.

Phytolyma nymphs prefer to attack young leaves. This not only ensures that good-quality food is available, but also the insect is able to complete development

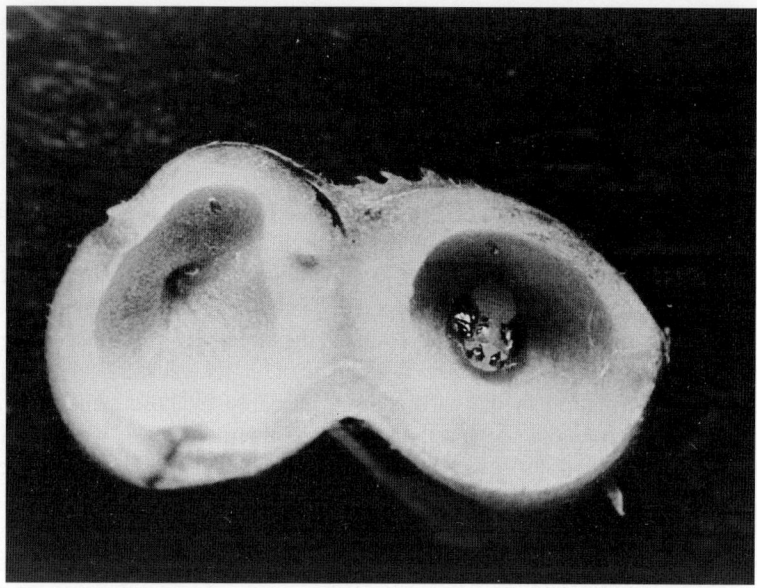

Figure 3.1 Nymph of *Phytolyma* sp. feeding inside of an opened gall formed on *Chlorophora excelsa*

before leaf fall (Cobbinah 1986). It is possible that, in addition to chemical factors, physical properties such as cuticle thickness may affect preference. Young leaves tend to have lower cuticle thickness than older leaves. The distribution of galls on leaves is also influenced by the surface of the leaf, region of the leaf, and size of the veins. White (1968) gives the total life cycle of *P. fusca* as 22 days, while Orr and Osei Nkrumah (1978) indicate that the life cycle of *P. lata* on young, vigorously growing *M. regia* is between 34 and 45 days.

Damage

Phytolyma spp. attack is more injurious in nurseries and young, vigorous plantations than in natural forests. At Mesewam, both galls and adults of *P. lata* are present throughout the year, although in low numbers between January and March and in September. When plant tissues are heavily infested (Plate 30), decay sets in rapidly. The shoots and leaves become a putrefying mass and the stem dies back as far as the woody tissue. Two fungal species, *Lasiodiplodia theobromae* and *Fusarium solani*, were isolated from branches of *M. excelsa* twigs and linked to the dieback (Apetorgbor et al. 2001). The plant is rarely killed by this initial attack and may send out auxiliary shoots, which are invariably attacked while still only buds. After sustained attack, however, even the healthiest seedling will have retarded or stunted bushy growth and may eventually die. Nursery plants and plantations in the first year of growth may suffer the greatest damage; 100% failures have been reported in various nurseries and plantations in Ghana.

Pest Management

Field tests in Nigeria by White (1968) using systemic insecticides gave promising results. Granules of Phorate (0,0-dimethyl-S-(ethylthic) methyl phosphorodithioate, 10% active ingredient), Dimethoate (0,0-dimethyl-S-(N-methyl carbamylmethyl) phosphorodithioate, 10% a.i), and Solvirex (0,0-diethyl-S-2-(ethylthioethyl-phospho-rodithioate), 5% a.i) applied as a soil dressing at the rate of 2.5–5.0 gm/seedling gave complete protection against *Phytolyma* attack for 12–15 weeks. The insecticides prevented gall formation in the early stages, probably by killing the first instar nymph as soon as it began sucking the sap. When the insecticides were applied to the plants already galled, the growth of the smaller galls appeared to cease when the nymphs were killed. The larger galls, however, continued to grow and in some cases even reached the point of dehiscence, revealing the dead insect inside.

Another field trial was conducted at Mesewam using five different insecticides: Bidrin, Unden, Aldrex 40, and Azodrin. When sprayed on gall tissues, all five insecticides resulted in significant mortality on *P. lata* nymphs after 24 h at 0.07% levels of concentration. Ripcord (a pyrethriod) and Aldrex 40 (an organochlorine) were not effective against *P. lata* adults except at concentrations above 0.3%. The systemic insecticides, Azodrin, Bidrin, and Unden, gave good control of adult *P. lata*, although the organophosphates Azodrin and Bidrin were more effective than the carbamate Unden at equivalent dosage rates (Cobbinah 1983).

Cobbinah (1988) proposed an integrated control program for *P. lata* based on data collected at Mesewam and other parts of Ghana (Table 3.1). Insecticides should be applied before April, July, and November to break the population peaks occurring at this time of the year. In March, when most of the trees are bearing young leaves (regeneration stage), it is necessary to apply chemical insecticides to all parts of the tree. In June and October, however, insecticides should be applied to the upper crown.

Integrated Control of *Phytolyma* spp.

Beginning in the early 1990s and throughout the decade, sustained financial support from the International Tropical Timber organization (ITTO) made it possible for a full-scale study of the *Phytolyma* problem in West Africa. The studies were coordinated by the Forestry Research Institute of Ghana (FORIG), in cooperation with Northern Arizona University, USA, and involved a number of scientists and institutions in Africa, Europe, and North America (Cobbinah and Wagner 2000). The following pest management strategies were investigated: genetic resistance, cultural control, including companion and mixed-species planting, and biological control options.

Table 3.1 Strategies to be considered for integrated control of *Phytolyma lata*

Strategy	*Phytolyma lata* stage affected	Type of control
Egg parasitism by *Trichogramma* sp.	Egg	Biological
Predation by mantid	First instar crawlers and to a lesser degree exposed adults	Biological
Boring of gall tissues followed by para-sitization of nymphs by hymenop-terous parasites: (a) Encyrtidae; (b) Eulophidae		Biological
Host selection: preliminary trials should ascertain which of the two major species (*Milicia excelsa* or *M. regia*) is more susceptible to the local *Phytolyma* complex	Affects the insect species as a whole through its oviposition preference, rate of establish-ment, fecundity, and other growth characteristics	Varietal resistance
Mixed planting	Adults	Cultural
Insecticide application: Azodrin, Bidrin, Unden	All stages	Chemical

Host Resistance

Progeny and provenance trials have been conducted for a wide range of seed sources from across the range of *Milicia* species in Africa (Plate 31). Beside Ghana, seeds were obtained from Côte d'Ivoire, Sierra Leone, Cameroun, and Tanzania. Based on the phenotypic expressions of the individuals screened three levels of resistance to *Phytolyma* were observed namely, resistant, tolerant, and susceptible progenies (Ofori and Cobbinah 2007). Classification of *M. excelsa* into resistant (antibiosis) or susceptible progenies was based largely on the ratio of incidence of small galls to large galls following *P. lata* attack (Plate 32). Small galls were less than 3 mm in size, hard and rarely opened to release adult *P. lata* (resistant). On the other hand, large galls were between 3 and 10 mm in diameter, fleshy, and frequently burst open to release adult psyllids for reinfestation to occur (*susceptible*) (Cobbinah and Wagner 1995). In between the two extremes occurs the *tolerant* progenies in which the plant usually recovers after attack to economically acceptable levels. Although the initial classifications were based primarily on phenotypic characters, DNA analyses have revealed that the observed differences are genetically controlled (Ofori 2001; Ofori et al. 2000). Some highly tolerant genotypes have been discovered that grew well despite heavy population pressure by psyllids (Plate 33).

Silvicultural Control

Companion- and mixed-species planting techniques have shown some promise for managing *Phytoyma* in plantations. At Mesewam, *M. excelsa* growing under the shade of the nitrogen-fixing *Gliridia sepium* had fewer galls than *M. excelsa* grow-ing alone (Wagner et al. 2000; Plate 34). Although no differences were observed in height growth and number of leaves between plots, significantly fewer galls were observed on *Milicia* in companion plots during two sampling periods. The positive effects of shading and nitrogen were demonstrated experimentally (Wagner et al. 1996, 2000).

After 1 year, *Milicia* seedlings growing in 82% shade were nearly 50% taller than seedlings growing in 57% shade, and 100% taller than trees grown in full sun-light. There were generally fewer galls on seedlings growing under shade though leaf production was higher on seedlings in full sunlight. Two fertilizer treatments, (N-P-K) 15-15-15 and 10-4-22, significantly affected *Milicia* height when compared to the control. The higher nitrogen content fertilizer (15-15-15) resulted in signifi-cantly fewer branch diebacks. Diebacks are positively related to the number of galls (Cobbinah and Wagner 1995).

Attempts were also made to evaluate the impact of *Phytolyma* on *Milicia* when planted in mixtures with other timber species (Plate 35). Various trials were car-ried out at Mesewam and Bobiri Forest Reserve (BFR) with *M. excelsa* at 11%, 25%, 50%, and 100% densities in combination with at least one of the following timber species, *Albizia adianthifolia, Khaya ivorensis, Terminalia superba, Ceiba pentandra*, and *Tectona grandis*. Overall results showed lower gall numbers in mixtures with less *Milicia* and also lower gall numbers initially at the forest (BFR) than at the nursery site (Mesewam) (Nichols et al. 1999). After almost a decade, shade and site were still the major factors responsible for reduced gall incidence on *M. excelsa*, although excessive shading caused significant reduction in growth (Bosu et al. 2006).

Management recommendations for *P. lata* (Wagner et al. 2000), are as follows:

a. Planting *Milicia* beneath the partial shade of another species should be encouraged because it will grow faster and suffer less damage from *P. lata. Milicia* may be planted in the existing shade within a natural forest or alongside a fast-growing species that quickly overtops it.
b. Planting *Milicia* at a density of about 10% is recommended. Although such an approach requires additional thinning costs, the added value should easily justify the costs.
c. No specific companion species is recommended at this time, but choosing a fast growing, nitrogen-fixing species seems appropriate. The planting of at least four to five species in a mixed plantation is suggested.
d. It is recommended that small- to medium-scale operational mixed species plan-tations are established using genetically improved *Milicia*. Such plantations should be carefully monitored to determine how broadly applicable the results of this study are across the natural range of *Milicia* in Ghana. Vegetative propa-gation techniques using leaf cuttings (Plate 36) and tissue culture (Plate 37) have been developed to support the propagation of genetically improved *Milicia*. Given the large size that *Milicia* reaches (Plate 38), it will be decades before the full impact of genetic improvement can be assessed.

Natural Control Agents

Endemic natural enemies of *Phytolyma* have been documented (White 1964, 1968; Bosu et al. 2000), but it was not until the mid-1990s that a comprehensive evaluation was carried out (Bosu 1999). The natural enemies' complex recorded on *Phytolyma* at Mesewam comprised four parasitoid species, about a dozen predators, and several pathogenic fungi (*Fusarium, Aspergillums, Cladosporium*, and *Penicillium* spp.). The parasitoid species were *Psyllaephagus phytolymae* (Hymenoptera: Encyrtidae), *Aprostocetus roseveari, A. salebrosus*, and *A. trichionotus* (Hymenoptera: Eulophidae). Although parasitoids occurred all year round at Mesewam parasitism levels were quite low, and averaged between 4% and 16%. (Bosu et al. 2004). This level of parasitism is not enough to bring *P. lata* population below economic thresholds. Parasitoid population increases during the dry season and decreases during the wet season. During a 1-year evaluation at Mesewam (July 1996–June 1997), the lowest parasitoid population occurred during the month of October (9.6 parasitoids/1,000 galls) while the peak population occurred in February (82.2 parasitoids/1,000 galls). In general, parasitism increased as gall size decreased. Also, parasitism was higher on resistant plants than on susceptible or tolerant plants. Based on these observations, it has been recommended that an effective biological control of *Phytolyma* could be achieved in plantations planted with resistant *M. excelsa* genotypes.

Predation of the psyllids is generally low because for most of its life span *P. lata* is concealed in a gall. This cryptic habit of *P. lata* perhaps explains why very little attention has been paid to the study of predators. The most important predators recorded at Mesewam are mantids, *Sphondromantis lineola* and *Pseudocreobotra ocellata*. Also important were the assassin bugs, *Rhinocoris* spp., *R. bicolor, R. rapax, R. camelita*, and *Phonoctonus fasciatus* (Bosu et al. 2003). While the list of predators looks impressive their polyphagous habits make their impact on the *P. lata* almost insignificant. Although mantids have been shown to consume a large number of psyllids in the field, and confirmed by laboratory bioassays, field cage tests have shown that mantids, and indeed predators in general, are not reliable biocontrol agents because of their polyphagous feeding habits. Polyphagous insects feed on a number of hosts depending on the availability and accessibility of the host. This lack of host specificity of otherwise very useful *P. lata* predators imposes a major limitation to their usefulness in a pest control program. However, mantid predators especially may be important in complementing the role of parasitoids in regulating *Phytolyma* populations.

Diclidophlebia spp. (Hemiptera: Psyllidae)

Diclidophlebia eastopi Vondracek and *D. harrisoni* Osisanya are two important sap-sucking pests of Ghana forests. *Diclidophlebia eastopi* is popularly known among foresters as "white aphids" because of the large quantities of translucent wax secreted by the nymph. This psyllid is a common pest of *Triplochiton scleroxylon* in Nigeria, Ghana, and the Côte d'Ivoire. Eidt (1963) was the first to recognize this insect's importance in Nigeria. Later work in Nigeria (Osisanya 1968) revealed that both *D. eastopi* and *D. harrisoni* are involved. *D. eastopi* was unknown in Ghana until the advent of reforestation in the 1970s.

Description and Life History

The eggs of *D. eastopi* and *D. harrisoni* are similar in appearance and can be distinguished by where they are deposited on the leaf tissue. Eggs are pale white, cylindrical in shape, and approximately 0.4 mm in length. When the eggs are laid, the pedicel is driven through the epidermis to enable the egg to obtain water as well as support. *D. eastopi* usually oviposits in batches of 3–46 eggs, but sometimes single, regular intervals, close together in rows. These single eggs are usually laid at the beginning of the oviposition period before the first batch of eggs is deposited.

Diclidophlebia eastopi generally oviposits along the principal veins, especially in the area proximal to the petiole of mature leaves. Eggs are rarely deposited on young leaves. It is probable that *D. eastopi* oviposits along the veins near the petiole because this is where soluble nitrogenous nutrients are concentrated and where feeding occurs. *D. harrisoni*, on the other hand, prefers the leaf edges of only the young leaves and 90–95% of eggs are deposited on leaf margins. This habit is related to the preference of the *D. harrisoni* nymphs for feeding along the leaf edges. The eggs hatch within 8 days and tiny yellowish nymphs emerge. The nymphs live in colonies of 10–20 individuals and pass through five instar stages. Adults of *D. eastopi* are small insect-resembling minute cicadas. They are approximately 2.5 mm in length with a wingspan of approximately 5 mm. The opaque, metallic, or blackish wings are held roof-like over the body when the insect is at rest (Figure 3.2). The head is relatively wide with large lateral compound eyes, and a pair of filiform antennae with fine apical bristles. Three ocelli are present: two close to the compound eyes and one near the anterior end of the head. The adult of *D. harrisoni* has yellowish to brown coloration. The life cycle is completed within 3–4 weeks.

Damage

Attack in the nursery can be very serious and entire plots may be destroyed. In established plants, attack results in dieback, stunted growth, and copious branching of the stems. The sucking insects empty the contents of the leaf cells, removing the green color and resulting in chlorosis. Feeding by nymphs may ultimately result in complete defoliation and death of the plant (Roberts 1969). Young trees in their first year in transplant beds are very susceptible to attack (Plate 39). The periphery of the leaves is most seriously affected; they curl, roll, and later turn yellow or brown (Plate 40, 41).

Nymphs are first evident at the end of the rainy season and are present in the greatest numbers in the early dry season. Severe damage to *T. scleroxylon* seedlings

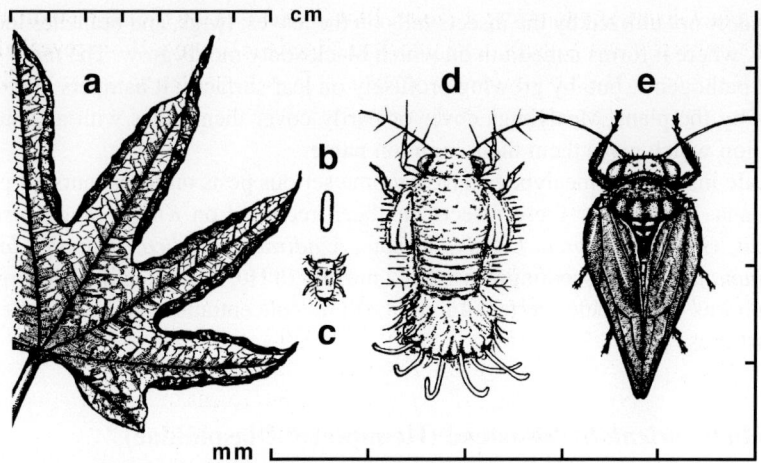

Figure 3.2 The psyllid *Diclidophlebia eastopi* is an important pest of *Triplochiton scleroxylon*: (a)wawa leaf with damage; (b)egg; (c)first instar nymph; (d)mature nymph; (e)adult. (From Kudler 1968; reprinted with permission of FPRI, Kumasi.)

was reported at Subri South-West Forest Reserve nursery in May when damage reached 75%. A similar outbreak was reported in the Esen Epam Forest Reserve in a 3-year-old *T. scleroxylon* taungya plantation. *Diclidophlebia* spp. has been found on flowering and fruit-forming sprouts of mature trees, as well as on leaves.

Pest Management

In Nigeria, chemical control is effectively obtained with two applications of 0.1% Malathion emulsion, 3–4 weeks apart. The insecticide Perfeckthion applied at 0.05–0.1% may also be used to affect some measure of control. Culturally, starting seedlings in the early rainy season can be an advantage because the seedlings are well established before the advent of the dry season when populations of the psyllids reach a peak.

Scale Insects (Hemiptera: Coccidae)

Scale insects are so called because many species secrete a scale-like wax coating over their back. They are usually small, peculiar creatures and do not look like insects. These insects cause general devitalization and death of the infested plant parts by extracting sap. Different species attack different parts of the hosts, but most infest young tissues, apical shoots, and particularly newly opened leaves causing premature leaf fall.

Scale insects, and a closely related family, Pseudoccocidae (mealybugs), frequently produce large amounts of honeydew, which attracts nectar-feeding insects. The honeydew not utilized by the insects falls on the leaves, twigs, and branches located below, where it forms a medium on which black sooty molds grow. The mold itself is not pathogenic, but, by growing profusely on leaf surfaces, it hampers photosynthesis by the plant. Mealybugs cover or partly cover themselves with a flour-like secretion which gives them their common name.

Scale insects and mealybugs have become serious pests on forest nursery plants in Ghana. Damage by these insects has been recorded on *Khaya sengalensis, T. grandis, Gmelina arborea, Eucalyptus* spp., *Azadirachita indica, Cedrela odorata,* and *Samanea saman* seedlings raised at a nursery in Ho, Volta Region. The organophosphorus insecticide Perfeckthion at 0.1% concentration is found effective against those pests.

Aonidiella orientalis Newstead (Hemiptera: Diaspididae)

The oriental yellow scale insect *Aonidiella orientalis* has emerged as a major pest of neem (*Azadiratcha indica*) in the dry savannah areas of West Africa (Figure 3.3). It is particularly serious in countries within the Lake Chad basin which includes Niger, Nigeria, Chad, and Cameroun (Lale 1998). The emergence of this insect in Africa is a classical example of an exotic pest invading the continent. It is believed to have been introduced from India, Southeast Asia, or possibly China. The first major record of *A. orientalis* outbreak was that on neem in northern Cameroun in 1985. Since then, the scale has spread to over 1 million square kilometers. Although neem is the principal host species it is reported that the scale also attacks about 30 other trees, ornamentals or fruits trees (Figures 3.4 and 3.5).

Description and Life History

Scale cover of adult female is 1.5–2.6 mm in diameter, circular to oval, flat, off-white to pale brown or yellow, with yellow to dark brown exuviae positioned more or less centrally. Beneath the scale cover, the adult female insect is pyriform initially, expanding with maturity to subcircular and becoming moderately sclerotized around margins. Scale cover of male similar in color to the female but smaller, elongate oval, with subterminal yellow exuviae, the adult male is winged (Ghauri 1962).

A. orientalis life history has not been studied in Africa, however, in Asia reproduction is usually by sexual means, although parthenogenetic and viviparous forms of reproduction have also been observed (Farid 1994). It has been speculated that adult females produce species-specific sex pheromones that attracts the males during reproduction (Khalaf and Sokhansani 1994). Females produce hundreds of eggs and protect them with a waxy covering. Development from the first instar or crawler stage to the adult took an average of 19.5 days in males, while it took females an average of 44.2 days from the crawler stage to the production of the next generation of crawlers. Glover (1933) reported three generations per year in India.

Figure 3.3 The oriental yellow scale insect *Aonidiella orientalis* on neem leaves. (Reproduced with permission of Maisharous Abdou. Direction de l'Environment, Niamey, Niger.)

Figure 3.4 Pruning neem trees infested by the oriental yellow scale insect *Aonidiella orientalis* in Niamey, Niger. (Reproduced with permission of Maisharous Abdou. Direction de l'Environment, Niamey, Niger.)

Figure 3.5 Predatory ladybird beetles (Coccinellidae) feeding on oriental yellow scale insects

Damage

Attack is followed by premature browning which frequently leads to death of the leaves on some or all of the branches of the affected tree (Lale 1998). Affected leaves occasionally remain yellowish-green in color but may disintegrate readily when rubbed in between the fingers. Infested leaves have spotted or necrotic patches, while typical dieback of small twigs and branches may occur. *A. orientalis* attack also leads to premature fruit drop. Trees 10–15 years or older are more susceptible to attack than younger trees, however, low environmental moisture is a major predisposing factor to outbreaks. There are two peak seasons in West Africa which occur during the dry season, March–May, and the wet season, July–September, respectively.

Pest Management

Cultural and chemical control methods have been tried in the affected countries. Trees with infested branches were pruned (Figure 3.4) but in more severe cases whole trees were removed and incinerated at designated locations. Trees or branches removed before the onset of the rainy season coppiced more rapidly and had lower risk of reinfestation. However, this control strategy remains effective when affected localities or member countries participate fully and almost simultaneously.

Chemical control in Nigeria and Cameroun using organophosphates in heavily infested areas was only partially effective. The thick waxy protective scale serves as a strong barrier against contact insecticides. Systemic insecticides may be more

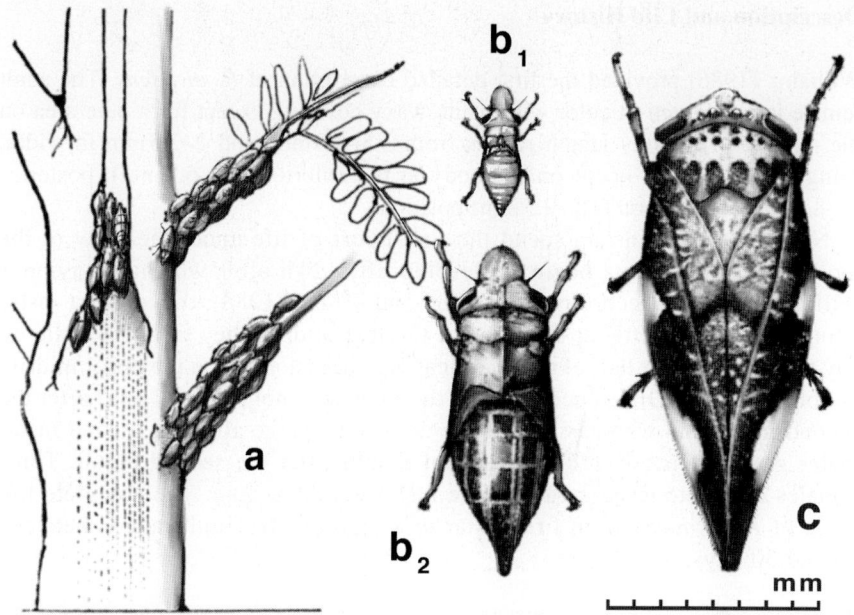

Figure 3.6 *Ptyelus grossus* (a)feeding nymphs in a colony'; (b$_1$, b$_2$)nymphs; (c)adult, a common and potentially serious pest of *Sesbania grandiflora*. (From Kudler 1970a–c; reprinted with permission of FPRI, Kumasi.)

effective but this will only work well during the wet season when there is enough water in the soil to transport the chemical up the stem.

Several important predators (Figure 3.6) and parasites have been found on *A. orientalis* in the Lake Chad Basin (Akanbi et al. 1990), which gives bright prospects for biological control of this insect. Notable among the predators is the ladybird beetle, *Chilocorus dislignia*. Other natural enemies recorded on *A. orientalis* are species of *Compriella, Tetrastichus, Aphytis, Cybocephalus*, and *Chilocorus*. In addition to all these, efforts are being made to introduce resistant neem varieties.

Rastrococcus invadens Williams (Hemiptera: Pseudococcidae)

The mango mealy bug (*Rastrococcus invadens*) is another example of an exotic invasive pest in Africa (Plate 42, 43). The insect was probably introduced into Ghana or Togo accidentally in the early 1980s. It is now very well established throughout West and Central Africa and seriously threatens production of mangoes and citrus in Togo, Benin, Côte d'Ivoire, Nigeria, and the Congo (William 1986). Until its invasion and subsequent study in West Africa, *R. invadens* was originally identified as *R. spinosum* (Robinson), a closely related species which infests fruit trees in Asia.

Description and Life History

Williams (1986) provided the first detailed description of *R. invadens*. The adult female is pale green in color with white waxy covering except for a bare area on the midline. The body length ranges from 3.5–4.0 mm, and 2–2.5 mm in width. Long wax filaments occur on the body at the anterior (3.3–6.0 mm), posterior (5–8.0 mm), and lateral (1.5–2.5 mm) positions.

Newly emerged instars spend the first hours of life under the body of the mother. Wax filaments begin to develop after 24 h after which the nymphs settle on the leaf. According to Willink and Moore (1988), though first instar nymphs prefer to settle at the sides of the leaf midrib, they can also settle on any part of the lower half of the underleaf, and occasionally on the upper midrib, petioles, or stem. Difference between the sexes was not apparent until after the second moult and whereas the female does not change after the second instar males go through two additional pupal moults after the second moult. Thus, females complete three moults while males complete four. The complete life cycle of *R. invadens* from first instar to a reproductive adult ranged between 45 and 50 days.

Damage

Mealy bugs usually attack the leaves of mangoes, citrus, or other fruit trees. However, petioles, flowers and fruits are also frequently attacked. The insect sucks sap from the plant and secretes honeydew, which deposits on the leaves or other parts of the plant. The honeydew provides a conducive substrate for growth of black sooty mould (Plate 43). At the beginning of the infestation, only the leaf base is affected. However, the entire leaf becomes covered with mealy bugs over time (Agounke et al. 1988) disrupting normal photosynthetic activities and thereby decreasing fruit production. Mealy bug infestation significantly reduces fruit production and causes substantial reduction in income to mango and fruit farmers throughout West Africa. In addition, mealy bug infestation of mango trees significantly reduces the esthetic and shade value of mango trees in communities.

Pest Management

The initial reaction of farmers to severe mealy bug infestations was to spray trees with large doses of pesticides or cut down affected trees. However, these chemical and mechanical control measures did not prevent the spread of the mealy bug. Biological control efforts using the parasitic wasps *Gyranusoidea tebygi* and *Anagyrus mangicola* (Hymenoptera: Encyrtdiae) have helped to considerably reduce the population of the mealy bug. Expected benefits of the biocontrol program run into millions of dollars (Bokonon-Ganta et al. 2002).

Other Sap-Sucking Insects

Two additional sap-sucking insects are worthy of note. *Ptyelus grossus* F. (Hemiptera: Cercopidae) is a common and potentially serious pest of *Sesbania grandiflora* (Kudler 1970a–c). This species tends to be abundant in the early dry season. Nymphs occur in colonies (1–3 per tree) of 20–40 individuals. The nymphs appear similar to others in this family and they produce the typical "spittle" for which this family receives its common name, "spittle bugs" (Figure 3.6).

A second sap-sucking insect of potential importance is the tree hopper *Otionotus* sp. (H: Membracidae) (Kudler 1978). This species is found in great numbers and is a potential threat to seedlings of *Terminalia ivorensis*. The nymphs and adults of this species are typical for the family: the adults have a large pronotum that covers the head (Figure 3.7). Some members of this family also can cause damage as a result of oviposition.

Table 3.2 lists the known sap-feeding insects of Ghana. The host plants and life history are given for each species listed.

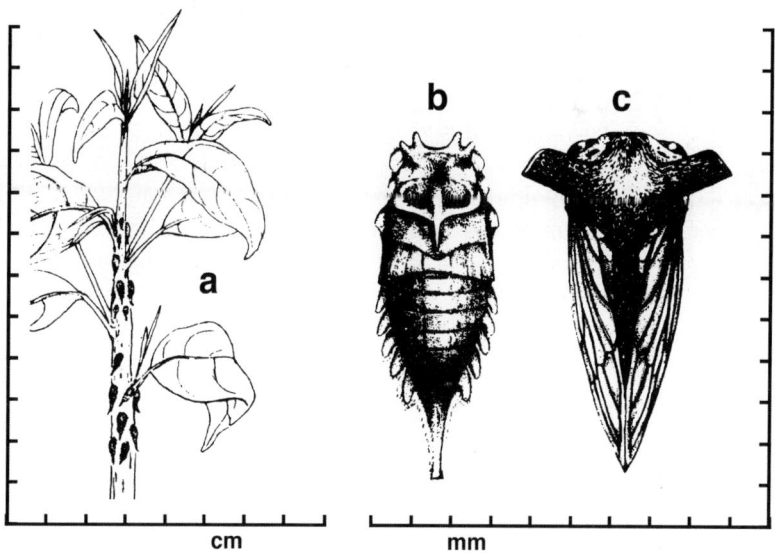

Figure 3.7 *Otionotus* sp. is a potentially serious pest to *Terminalia ivorensis* seedlings: (a)nymph colony; (b)nymph; (c)adult. (From Kudler 1978; reprinted with permission of the Forestry Science Institute, Prague, Czechoslovakia.)

Table 3.2 Sap-feeding insects of Ghanaian forests

Species	Order: family	Host plant	Life history/comments
Aulacaspis tubercularis	Hemiptera: Diaspididae	Mangifera indica (mango)	Sap-feeder
Coccus elongatus Signoret	Hemiptera: Coccidae	M. indica (mango)	Sap-feeder
C. hesperidum Linnaeus	Hemiptera: Coccidae	Acacia spp., Dalbergia spp.	Sap-feeder
C. viridis Green	Hemiptera: Coccidae	Hevea brasiliensis, Mangifera indica,	Sap-feeder
Cryptoflata unipunctata Oliv.	Hemiptera: Flatidae	Terminalia ivorensis	Nymphs suck sap from shoots
Diclidophlebia eastopi Vondracek	Hemiptera: Psyllidae	Triplochiton scleroxylon	Suck sap from young leaves of host
D. harrisoni Osisanya	Hemiptera: Psyllidae	T. scleroxylon	Suck sap from young leaves of host
Ferrisiana virgata Cockerell	Hemiptera: Pseudococcidae	Casuarina equisetifolia, Leucaena glauca, Terminalia superba	Sap-feeder
Otionotus spp.	Hemiptera: Membracidae	T. ivorensis	Nymphs suck sap from shoots
Phytolyma lata	Hemiptera: Psyllidae	Milicia regia	Sucks sap from leaves and twigs; several secondary insects and pest species have been found in association with Phytolyma galls at Pra-Anum Forest Reserve, including Ahasuerus adverna Walt.) (Cucujidae) Araecerus fasciculatus ODeG.) (Anthiribidae), Eublema sp. nr. aurantica Hamps. (Noctuidae), Cryptoblabes gnidiella Mill. (Tortricidae), Lobesia selopa Meyr. (Tortricidae), Pyrodercus sp. (Cosmopteridae)
P. fusca	Hemiptera: Psyllidae	Milicia excelsa	
P. tuberculata	Hemiptera: Psyllidae	M. regia	
Ptyelus grossus Fabricius	Hemiptera: Cercopidae	Sesbania grandiflora	Common on its host; brownish nymphs and adults that are most abundant at the beginning of the dry season; feeding results in extensive sap flow along the tree stem
Planococcoides njalensis Laing	Hemiptera: Pseudococcidae	Tectona grandis, T. ivorensis	Sap-feeder

P. citri Risso	Hemiptera: Pseudococcidae		Sap-feeder
P. kenyae LePelley	Hemiptera: Pseudococcidae		Sap-feeder
Pseudococcus adonidum Linnaeus	Hemiptera: Pseudococcidae		Sap-feeder
Pseudophacopteron zimmermanni Aulmann	Hemiptera: Psyllidae	*Khaya anthotheca, K. grandifolia, K. ivorensis, K. grandifolia*	Produces numerous galls (3–5 mm) in diameter on hosts
Stictococcus sjostedti Ckll.	Hemiptera: Coccidae	*Musanga cecropioides*	Found sucking sap from leaves of its host at Mpraeso (Northern Scarp)
Toxoptera aurantii Boyer de Fonscolombe	Hemiptera: Aphididae	Cacao	Sap-feeder
Triozamia lambourni Newstead	Hemiptera: Psyllidae	*Antiaris africana*	Common pest in rain forests; trees of all ages are attacked
Udinia faraguarsoni (Newest.)	Hemiptera: Diaspididae	*K. ivorensis*	Sucks sap from abaxial portion of leaves

Chapter 4
Wood Borers of Living Trees

Introduction

Stems of living trees provide food and shelter for many species of boring insects. Plants in various stages of development from very young seedlings to mature trees may be attacked. The damage caused by these insects results in the death of living stems and shoots and reduction in the wood quality due to boring and weakening of standing trees. In addition to these direct effects, borers may open the way for secondary attack by wood rotting or staining fungi, thus hastening decay.

Representatives of borers of living trees in Ghana can be found in two principal insect orders: the Lepidoptera (Cossidae, Noctuidae, Pyralidae, and Thyrididae) and Coleoptera (Bostrichidae, Cerambycidae, Curculionidae, Platypodidae, Scolytidae, and Tenebrionidae). Of these two orders, the greatest numbers of economically important pests occur in the Coleoptera. In addition to the taxonomic groupings mentioned above, borers can be classified based on: (a) condition of the host tree attacked; (b) range of host plants attacked; (c) developmental stage of the plant when attacked; and (d) portion of the tree inhabited by the insects.

Classifications of Wood Borers

Host Condition

Browne (1964) classified ambrosia beetle attack on living trees in tropical forests into five distinct categories. This classification is adopted here for all the insects covered in this chapter. The categories are: (1) attack of apparently healthy trees, for example, *Trachyostus ghanaensis* attack on wawa (*Triplochiton scleroxylon*); (2) sporadic mass attacks on healthy trees, such as the attack by *Doliopygus dubius* on ofram (*Terminalia superba*); (3) attack on tree species when their vigor is temporarily reduced by drought or unseasonable weather, such as *Bostrychoplites cylindricus* attack on osempe (*Cassia alata*); (4) attack on very unhealthy and dying trees, such as attack by *T. aterrimus* on esafufuo (*Celtis mildbraedii*); and (5) attack through wounds, which is normally of little importance in Ghana. Ambrosia beetles attacking felled logs is covered in a separate section.

Host Range

Borers can also be classified based on the number of plant species they utilize as food and/or shelter. One group uses a single plant or closely related plants as food and is referred to as monophagous. The emire shoot borer, *Tridesmodes ramiculata*, and the wawa borer, *T. ghanaensis* are monophagous. A second group includes insect species that use plants from a few genera or several closely related genera as hosts and are referred to as oligophagous; an example is the mahogany shoot borer, *Hypsipyla robusta*. A third group of borers, polyphagous, uses hosts belonging to many different plant families and orders; a notable example is *Apate monachus*.

Developmental Stage of Trees When Attacked

Borers of living trees can be divided into two groups based on the developmental stage of their host. Shoot borers, insects that bore into seedlings, saplings, and young plants, are of great economic importance because they attack trees when the greatest annual growth increments are being accrued and because their damage causes trees to fork (Plate 44). The mahogany shoot borer and the emire shoot borer are the most prominent members of this group in Ghana. A characteristic of shoot borer attack is extensive lateral bud development (Plate 45). In some cases terminal growth is prevented, and twig breakage and mortality may result. The second group of insects attack mature trees and includes representatives of several beetle families. Some, for example, *A. monachus*, attack healthy trees, but, in general, weakened or severely damaged hosts are preferred.

Feeding or Shelter Location

The classification system based on the part of the tree utilized by borers for feeding or shelter is more universally applied, has a wider application, and is used here. The categories include: (a) girdlers; (b) phloem borers; and (c) xylem (wood) borers. Girdlers include *Analeptes trifasciata*, which attacks members of the Bombacaceae, and *Tragocephala gorilla*, found on *Acalypha* spp. Both are members of the beetle family Cerambycidae. Insects in the phloem-boring group are represented mainly by the beetle family Scolytidae, and are commonly referred to as bark beetles. Most phloem borers are considered secondary insects because they infest unhealthy, suppressed, or dying trees. Examples of this group include *Xyloborus semiopacus, Polygraphus granulatus,* and *Xyloctonus scolytoides.*

The third group, also known as wood borers, is represented by the orders Lepidoptera and Coleoptera, although the latter group is most common in Ghana. Examples are *T. ramiculata, Orygomorpha mediofoveata,* and *Terastria reticulate* (Lepidoptera), and *Apate terebrans, Bostrychopsis tonsa,* and *Mallodon downesi* (Coleoptera). Wood borers often spend considerable time in the phloem before entering the xylem.

Pest Management

There is a number of management techniques used to lessen the damage caused by boring insects. Maintaining plant vigor with good silvicultural practices can greatly reduce pest damage. Removal and destruction of borer damaged trees or wind-thrown branches reduce the chances of subsequent attacks in the residual stand. For mature trees, rapid extraction and conversion of infested trees appear to be a good management practice. Selecting resistant plant varieties and planting primarily on suitable sites is also important. Throughout the tropics, economic considerations are increasingly leading to the establishment of forest plantations. Tropical foresters should, however, place a premium on establishing mixed plantations to minimize the risk of pest and disease problems. The biology of some of Ghana's economically important borers of living trees follows.

Lepidopterous Borers of Living Trees

Borers of living trees in the order Lepidoptera cause damage only in the larval stages. The most important lepidopterous borers are the mahogany shoot borer, emire shoot borer, and the opepe shoot borer *Orygmophora mediofoveata.* Another borer, *Eulophonotus obesus* (Karsh), has been recorded on wawa, *T. scleroxylon.*

Mahogany Shoot Borer, *Hypsipyla robusta* (Moore) (Lepidoptera: Pyralidae)

Mahogany species are some of the most valuable export tree species in West Africa. In both Ghana and Nigeria the mahogany shoot borer has created serious problems in mahogany plantations. The planting of mahogany has been almost completely abandoned because of the destruction caused by this insect. *Hypsipyla robusta* also attacks fruits, and bark. There is likely more than a single species of *Hypsipyla* in West Africa.

Description of Life History

The adult month is small, with a 3–5 cm wingspan. The first to third instar larvae are light brown to red in color. The last instar larva is distinctly greenish blue. The head and the pronotal and suranal plates are black. On each side of the body there are five rows of black spots (Plate 46, Figure 4.1). The complete life cycle in West Africa is not clearly known. Roberts (1968, 1969) recorded four larval instars. In Ghana, five or six larval instars have been recorded; most larvae pupated after the sixth instar. Atuahene and Souto (1983) also recorded five or six larval instars on artificial medium. The pupal stage lasts 8–9 days; there may be six to nine generations per year in the high forest zone. Roberts (1969) observed that *H. robusta* diapauses during the dry season and emerge as adults with the onset of the rains. Therefore, because the savannah has a longer dry season than the forest belt, there will be fewer generations (probably 3–5) of *H. robusta* in the savannah. Some of the uncertainty regarding life cycle may be related to the existence of more than a single species of *Hypsipyla* (Horak 2003).

Damage

Infested stands exhibit extensive lateral bud development because feeding by *H. robusta* interferes with the normal processes of apical dominance by the terminal

Figure 4.1 *Hypsipyla robusta* larva dorsal and lateral view

shoot (Plate 45, 47). *Hypsipyla robusta* larvae feed on the soft tissue inside the terminal stem. In heavy infestations, as many as 20–40 wounds may occur on the stem, resulting in heavy sap exudation (Plate 48). Attack is usually more severe on trees growing in full sunlight compared with shaded areas. Two main periods of infestation occur which correspond with the primary and secondary leaf flush of *Khaya senegalensis* and *K. ivorensis*.

Trees are first attacked in their second or third year. Generally, shoot borer attack weakens the tree and predisposes it to other insects and fungi. By constantly forcing the tree to direct a disproportionately large amount of its energy to terminal shoot growth, the attacks probable result in much less energy available for other requirements such as root growth. With repeated infestation, mortality can occur. Economic loss is usually due to reduced height growth and poorer form of infested trees.

In Ghana, *H. robusta* commonly attacks members of the genus *Khaya*. In Nigeria, this insect has been found attacking *Entandrophragma angolense, E. candollei, E. cylindricum, Khaya nyassica, K. senegalensis, Carapa grandiflora, C. procera, Lovoa trichilioides, Pseudocedrela kotshiyi*, and *Swietenia macrophylla* (Jones 1959a).

Pest Management

Attempt to control *H. robusta* larvae with systemic insecticides have been only partially successful. Considering the life history of this insect (i.e. short life cycle, multivoltine, overlapping generations), Wylie (2001) has suggested that it would require perhaps 3–5 years of continuous insecticide application in the field before adequate protection can be achieved. Even if this would be economically feasible, environmental considerations make this option quite inappropriate. Additionally, the cryptic habit of the larval stages (i.e. hidden or concealed in the shoots) makes contact insecticides ineffective. Control of adults with contact insecticides could be useful but in most countries population densities are highest during the wet season and insecticides easily lose their efficacy due to high humidity. Systemic insecticides are more promising, however, it has been noted that the more frequently used carbamate and organophosphate systemic insecticides are readily biodegradable in humid tropical environments. In general, insecticide application in nursery situations might be helpful in many ways; however, large-scale control of *Hypsipyla* in field situations is not practical.

Integrated Management of *Hypsipyla robusta*

Over the last 2–3 decades research efforts have increased substantially on the global scene to address the mahogany shoot borer problem. In Latin America, where *H. grandella* is the primary species, prevalent trials have been conducted on various new world mahoganies including *Swietenia, Cedrela*, and *Toona* spp. In Africa, India, Australia, and Southeast Asia, control attempts have been largely on *H. robusta*. However, the taxonomic and ecological relationships between these two species are such that any management experiences could readily be shared among various countries or regions. Experts generally agree that selecting resistance host and silvicultural control options have good prospects for managing *Hypsipyla* spp. (Flyod 2001).

Host Resistance

Variability in susceptibility to *Hypsipyla* spp. has been found between and within various species of the Meliaceae. In Ghana, provenance and progeny trials using 28 progenies of *Khaya* and *Entandrophragma* spp. showed substantial variability in growth and damage levels (Ofori et al. 2004, 2007; Opuni-Frimpong et al. 2004, 2005). Attack is generally more severe on *Khaya* than *Entandrophragma* spp., but damage to four *K. ivorensis* progenies were found to be within acceptable limits. Continuing research would determine whether or not the observed levels of resistance are under genetic control. Resistance to *H. grandella* in two Central American mahoganies, *C. odorata* and *S. macrophylla* has been clearly demonstrated (Newton et al. 1995, 1996, 1998, 1999; Watt et al. 2001).

Silvicultural Control

Control of *H. robusta* by silvicultural methods has also been successful. In Côte d'Ivoire, planting *Khaya* in low densities under shade of *Leucaena leucocephala* gave promising results during early years of growth (Brunck and Mallet 1993). Better protection and good growth were obtained by planting mahogany under natural forest canopy. Similar results have been obtained in Ghana. Growing *Khaya* under forest canopy shade prevented shoot borer attack but this resulted in significant loss in growth (Opuni-Frimpong et al. 2004). Companion and mixed species trials of *K. ivorensis* planted at a density of 25% or lower with *Albizia adianthifolia, T. superba, Cedrela odorata*, and *Azadirachta indica* have resulted in lower shoot borer damage compared to pure *Khaya* stands (Opuni-Frimpong et al. 2004; Bosu et al. unpublished). It has been shown that reproduction of *H. robusta* decreases on shaded leaves compared to unshaded leaves as a result of poor nutritional quality of the former (Mahroof et al. 2002). Also, pruning of all epicormic branches except the leading shoot can sometimes provide some protection but further studies need to be done (Cornelius 2001, Bosu unpublished).

Biological Control

Control attempts using natural enemies have been in progress for sometime. The natural enemy complex of *Hypsipyla* spp. includes parasitoids, predators, pathogens, and viruses (Hauxwell et al. 2001). Parasites of *H. robusta* have been found across West Africa (Table 4.1), however, these have not been evaluated (Sands and Murphy 2001). Attempts at classical biological control, whereby natural enemies are imported from one country to control the pest in another country have not succeeded. In spite of the failures, however, natural control still plays an important

Table 4.1 Parasitoids of *Hypsipyla robusta* in West Africa. (From Sands and Murphy (2001)

| Parasitoid | Family | Host stage | Country recorded | | |
			Côte d'Ivoire	Ghana	Nigeria
Hymenoptera					
Bracon sp. nr. *wellenburgensis* Wlkn	Braconidae	L	√		
Apalantes sp.	Braconidae	L	√		
Macrocentrius sp. (*linearis* group)	Braconidae	L		√	√
Microgaster sp.	Braconidae		√		
Protomicroplitis australis Wlkn	Braconidae	L	√		
Dioichogaster sp. (*spretus* group)	Braconidae	L		√	√
D. australis Wlkn	Braconidae	L	√		
Protomicroplitis sp.	Braconidae	L		√	√
Eucepsis sp.	Chalcidae	P		√	
Tetrastichus sp.	Eulophidae	P	√		
Eurytoma sp.	Eurytomidae	L		√	√
Diptera					
Cadurcia auratacauda (Curran)	Tachinidae	L	√	√	√
C. nr. *depressa* Villeneuve	Tachinidae	L	√		
Carcelia angulicornis	Tachinidae	L		√	
Ethyllina sp.	Tachinidae	L	√		
Parexorista amicula (Mesnil)	Tachinidae	L		√	√

L = larvae, P = pupae

role in the regulation of shoot borer populations, and might be especially useful if used in combination with other control strategies.

Emire Shoot Borer, *Tridesmodes ramiculata* Warr (Lepidoptera: Thyrididae)

T. ivorensis (emire) has long been recognized as a valuable species in the timber export trade. Due to initial plantation success, emire has been included in Ghana's forest improvement program. As a result, much information is available on insects attacking *T. ivorensis*. One of the most important pests is a shoot borer attacking nursery and plantation stock. The pest was identified in Nigeria as *T. ramiculata* Warr (Lepidoptera: Thyrididae). This same species was also found attacking emire in Ghana.

Description and Life History

Adult *T. ramiculata* moths have a wingspread of 18–22 mm (Figure 4.4). Wings are whitish and transversely speckled, with dark brownish irregular lines. The number of lines varies, but normally there is one distinct curved line near the outer margin

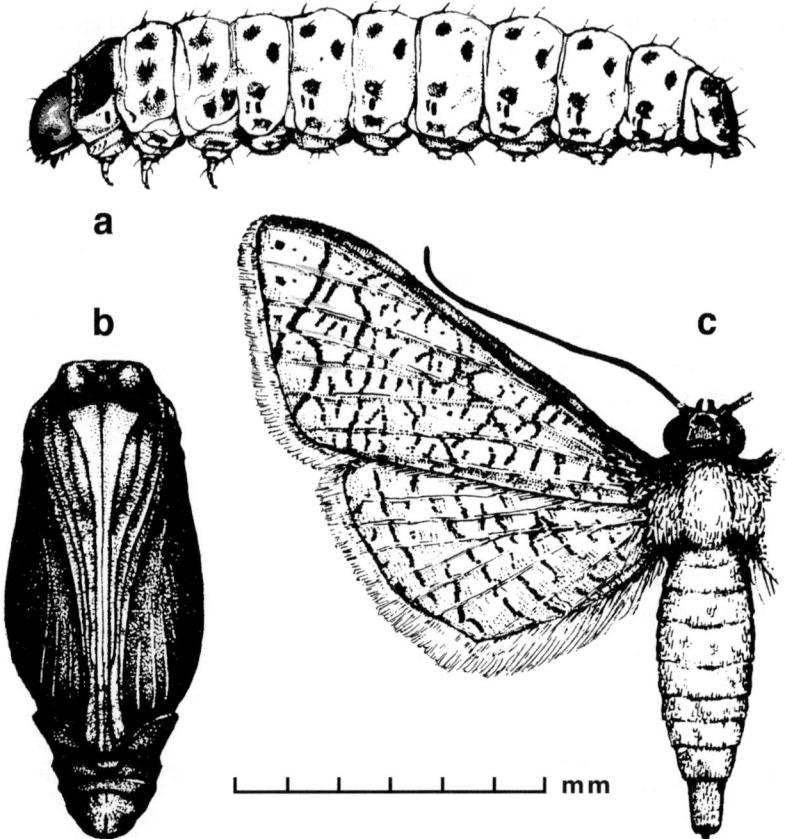

Figure 4.2 (a) Larva, (b) pupa, and (c) adult of *Tridesmodes ramiculata* that is important pest of emire. (From Kudler 1971; reprinted with permission of FPRI, Kumasi.)

of the forewing. Beyond this line, two or more parallel dark spots are present. The body is light brownish; tarsal segments have black markings; antennae are filiform and brownish (Figure 4.2).

Eggs are opaque yellowish, oval, and approximately 0.25 mm long. The first instar larvae are creamy white and have been seen penetrating the top of young shoots. The larvae consume the tissue within the shoot, pushing hard brown frass out of the entrance hole as they feed.

As the caterpillars increase in size, they tunnel from small shoots into the internodal region of larger shoots. The feeding tunnel is 4–8 mm in diameter and reaches a maximum length of 50 mm. Later instar larvae are creamy white to light brown or sometimes pale reddish to faintly bluish. The head and prothoracic shield are dark, usually orange-brown, in color; distinctive dark brown spots with setae occur on each segment (Figure 4.2). Mature larvae are approximately 15 mm in length.

Larvae pupate in the feeding tunnels for a period of 20–24 days. Pupae are dark brown and approximately 7–9 mm in length (Kudler 1971) (Figure 4.2). The life cycle is completed in 1–3 months, with many overlapping generations each year.

Damage

Tridesmodes ramiculata prefers actively growing shoots. Larvae feeding normally results in terminal shoot mortality and lateral bud development. This is very characteristic of shoot borer attack. If the lateral branches grow, the tree can overcome the attack; however, height growth is retarded. Frequently the branches are attacked simultaneously. The incidence of *T. ramiculata* damage varies with locality. In the Neung Forest Reserve, in southwestern Ghana, approximately 62% of the 2–3-year-old emire were attacked (Kudler 1971); in other areas, damage is minimal.

Pest Management

Newly hatched first instar larvae are susceptible to several predators and parasite. Lizards prey on the larvae, while a muscid fly, *Mydaea superba* Stein, feeds on both larvae and pupae. Little is known about the impact of these control agents on population levels of *T. ramiculata*. Chemical control trials by Kudler (1971) indicate that a 1% solution of Bidrin brushed on affected parts is effective in controlling damage. Because of the overlapping generations, this type of control is effective only if repeated frequently. Therefore, this control measure is practical only in the nursery. In many cases where attack is not heavy, vigorous plants outgrow the susceptible stage in a few years.

Eulophonotus obesus (Lepidoptera: Cossidae)

This moth was observed to cause considerable damage to saplings of *T. scleroxylon* in plantations within the Aboma Forest Reserve during the early 1970s. It is also recorded on the same host plant in Côte d'Ivoire and Nigeria. The moth has an average body length of about 3.0 cm and has transparent wings. The damage by the larvae is conspicuous and consists of deep galleries that run for many centimeters in the wood. Evidence of attack is a brownish sugary sap flow on the tree bole accompanied by wood frass pushed from the larvae tunnels. Damage may be observed throughout the year but is more common during the drier months of January–April. Entwistle (1963) recorded a closely related species *E. myrmyleon* on *Cola nitida* in Ghana and Sierra Leone. A survey conducted in a 291 even-aged stand of *T. scleroxylon* in Nigeria recorded attacks on 37% or 12.7% of the trees, with the highest incidence of attack being on the 62.5–75 cm girth range (Ashiru and Momodu 1981).

Good sanitation in the plantation may help to reduce insect numbers; however, in Ghana sawdust mixed with 0.3% Aldrex 40 when placed in the entry holes was found to be effective. Cotton wool soaked in Vetox 20 at the rate of 0.3–1% solution was recommended in a similar manner in Nigeria. Also, a combination of Furadan, Thimet (144 g/tree) or a combination of the two plus adequate weed maintenance provided good protection (Ashiru and Momodu 1986).

Orygmophora mediofoveata Hamps (Lepidoptera: Noctuidae)

Insect borers of *Nauclea diderrichii* (opepe/kusia) were observed in Nigeria as far back as the 1930s. *N. diderrichii* is an an evergreen tropical hardwood species belonging to the family Rubiaceae. It is known in the international timber trade as Opepe and locally in Ghana as kusia. However, the identity of the borer was not discovered until 1962 when Eidt (1965a) reared the moth for the very first time.

Description and Life History

The life history of *O. mediofoveata* is generally unknown, in particular the egg stage and the first instar larvae (Eidt 1965b). The larvae which attack and damage the plants are grub-like and morphologically quite unusual. The larva is short and stout; the head is partly withdrawn into the prothorax and the legs and prolegs are well developed. Fully grown larvae of the ultimate instar are about 14 mm long, and average head width is about 1.5 mm. Early instar larvae are translucent and appear greenish because of the plant tissue in the gut. Ultimate instar larvae are deep red on the dorsum but remain green on the venter. They infest the terminal shoots, boring in the last two or three internodes, and preferring the more apical shoot. They do not girdle the shoots, but bore in the pith and produce galleries several inches long. In the case of tunneling, the larvae eject dark brown frass which accummulates in the leaf axils. This tunneling can reveal their presence. Pupation occurs within the galleries although there is no cocoon. The pupal period lasts about 3 weeks. The length of a generation is unknown but has been estimated to be about 3 or 4 months. In Nigeria, moths from field-collected shoots emerged in the months of April, May, June, September, November, and December, indicating overlapping generations.

Damage

Damage is principally through stunting of the trees in nursery transplant beds, and is rare in seed beds. Attack results in the rapid formation of a callus tissue over the injured parts, and may lead to mortality in heavy multiple attacks. Parry (1956) noted that "attack by *Orygmophora mediofoveata* was not severe enough to discourage use of the tree in pure plantations in Nigeria, but in Ghana, opepe is so badly damaged that it is an unreasonable risk as a plantation crop." Record of infestation from other countries across Africa or elsewhere is unavailable but anecdotal evidences indicate that the borer is partly responsible for the lack of large scale *N. diderrichii* plantations in Ghana and the subregion. A recent survey of pests in the Ashanti Region of Ghana revealed significant levels of shoot borer infestations of *Nauclea* in nurseries and plantations (Bosu et al. 2004). Infestation levels of 30–80% were recorded in many new tuangya plantations in Nigeria during the early 1960s. In addition, unmistakable signs of old attacks were also found in older plantations. Recently in Ghana, between 60% and 80% of potted seedlings at the Fumesua and Mesewam nurseries of the Forestry Research Institute were destroyed by shoot borer attack, while about 50% of young plants in an experimenatal plot at the Pra-Anum Forest Reserve at Amantia were also infested.

Management

Eidt (1965a) reported the occurrence of three species of endoparasites on *O. mediofoveata* namely *Macrocentrus* sp. nr. *delicatus* Cresson (Hymenoptera: Braconidae), *Orgilus* sp. poss. *apostolicus* Turner (Braconidae), and *Pristomrus* sp. (Hymenoptera: Ichneumonidae). Chemical control has not been investigated but control with a contact insecticide directed at the first instar larvae between the time they hatch and when they enter the shoot has been recommended. Studies on genetic resistance and silvicultural control practices have recently commenced in Ghana.

Coleopterous Borers of Living Trees

This group contains the most important borers of living trees in Ghana.

The Polyphagous Beetles, *Apate monachus* and *A. terebrans* (Coleoptera: Bostrichidae)

The polyphagous beetles, *A. monachus* and *A. terebrans* are major forest pests in Ghana. These insects attack a wide variety of economic species in the country. The members of the family Bostrichidae are commonly referred to as branch and twig borers (Borror and Delong 1964). The adult beetles are elongated and somewhat cylindrical; the head is bent down and scarcely visible from above (Figures 4.3 and 4.4)

Description and Life History

The entrance tunnel of both species enters the xylem at an angle to a depth of approximately 1 cm (Thompson 1963). The oviposition tunnel follows the wood

Figure 4.3 *Apate terebrans* top view (body length 2.7 cm.)

grain. In standing trees the oviposition tunnels runs from the point of entry upward
and may reach 30 cm. The larval tunnels generally follow the wood grain and reach
a length of approximately 40 cm and a maximum width of 1.3 cm. They generally
exploit the entire available depth of the sapwood, but may also occur in the heartwood. The frass is packed very tightly in the tunnel.

Pupation occurs in the sapwood in an oval chamber which is a continuation of
the larval tunnel. The pupal period is approximately 3 weeks. The adult remains in

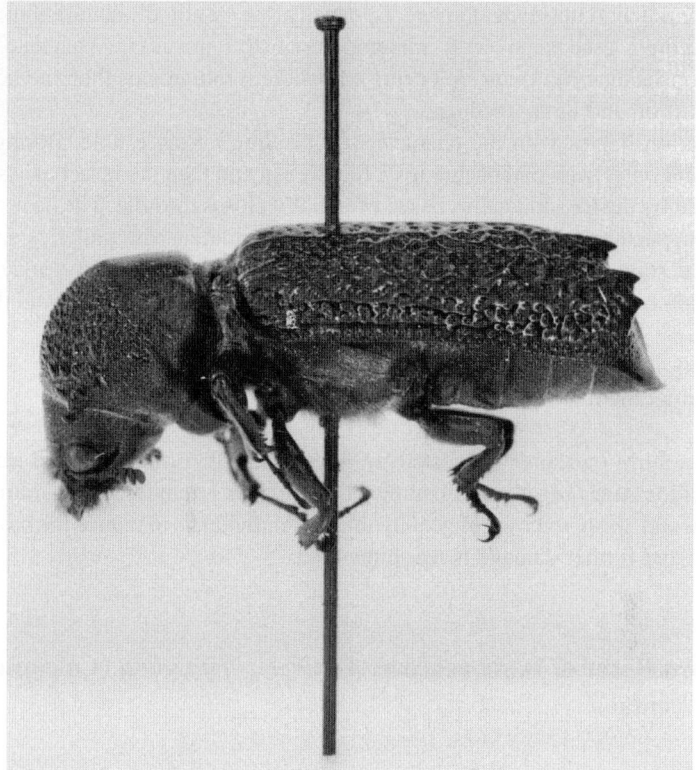

Figure 4.4 Side view of *Apate terebrans*. Note the head is pointed downward, which is typical for the family Bostrichidae

the pupal chamber for 1–2 weeks before boring a circular exit hole. In one infestation, the duration of the life cycle was 5 months (Thompson 1963), but this period may be prolonged under some conditions.

Damage

Damage caused by *A. monachus* and *A. terebrans* occurs in most vegetation zones of Ghana, but is especially evident in young plantations and taungya farms. Heavy infestation of *Eucalyptus polycarpa* by *A. monachus* and *A. terebrans* resulted in damage to 38.5% of a plantation in Sege. These insect pests attack both indigenous and introduced species, including *T. scleroxylon, T. ivorensis, Tectona grandis, E. polycarpa, Melia composita*, and several members of the *Khaya* genus (Atuahene 1976). They also attack *Sesbania grandiflora, Azadirachta indica*, and *Pithecellobium dulce*. Another less common species, *Apate indistincta* Murr, has been recorded attacking seedlings of *Pericopsis elata* (Atuahene 1976).

Economically important damage is usually the result of maturation feeding (adult feeding associated with maturation of the gonads) (Atuahene 1976). Maturation feeding by *Apate* sp. occurs on different individuals than the host used for oviposition and larval feeding.

Infestation begins with the characteristic slanting entrance hole. Boring in the phloem and outer sapwood occurs for a few weeks and then the insect exists. Gums are exuded by the tree and callus tissue eventually closes the hole. The heartwood of heavily infested trees such as *T. ivorensis* may be riddled with galleries and large patches of stain that render the tree valueless, although externally it may appear sound. This type of damage also increases the tree's susceptibility to wind damage.

Pest Management

Possible control measures suggested by Atuahene (1976) include good sanitation on plantation sites before outplanting. This involves removing and burning all infested wood. Also, removal of heavily infested individuals to reduce the probability of spread and further damage is recommended.

Longhorn Borer of Bombacaceae, *Analeptes trifasciata* (Coleoptera: Cerambycidae)

Longhorn borer of Bombacaceae, *A. trifasciata* E., is typically a savannah insect. It may, however, be found in relative abundance in forest edges where members of Bombacaceae grow (Atuahene 1975). The insect is common in the northern Ghana where severe attacks on *Ceiba pentandra* (Plate 49) have been recorded, but the same insect has also been found attacking *Eucalyptus* species in southern Ghana (Atuahene 1975).

Analeptes trifasciata attacks members of the Bombacaceae family, including *Bombax costatum, Adansonia digitata*, and *C. pentandra*. It also attacks *Eucalyptus tereticornis, E. alba*, and *T. grandis*. In Nigeria, *A. trifasciata* is known to feed and breed in *Anacardium occidentale* (Browne 1968). In West Africa, the insect has been recorded in Sierra Leone, Benin, Nigeria (Duffy 1957), and the Côte d'Ivoire (Lepesme 1953).

Description and Life History

The insect life history is known on Bombacaceae, but not yet clearly elucidated on *Eucalyptus*. After selecting a suitable host tree, the adult beetle feed on the bark and underlying wood. The adult has the typical appearance of cerambycids (Plate 50). That is, the antennae generally lie flat along the back and extend beyond the abdomen. The adult is approximately 5–6 cm in length and black with orange bands crossing the elytra. The branches of *B. costatum* that are attacked are generally 5–7 cm in diameter. In the case of *C. pentandra*, trees between 1 and 2 years with a diameter range of 10–20 cm are commonly attacked. After an initial period the beetles pair off; the male begins girdling the tree and is joined later by the females. In approximately 1–2 weeks, girdling of the tree is complete and the insects begin intermittent feeding above the girdled branch (Plate 51). Mating eventually occurs, interrupted by periods of feeding.

The mated female lays eggs singly above the girdle. The eggs are laid in small holes at the base of the large spines characteristic of these tree species (Figure 4.5). Each hole contains one egg and is covered with a papery capsule. When the larvae hatch, they feed beneath the back of the girdled stem. Large numbers of larvae can reduce the stem to a hollow cylinder packed with frass and the typical coarse wood fibers associated with longhorn beetles. The last instar and pupae have been found in May, June, and November through January, indicating that there may be two generations per year in Ghana.

Damage

The damage caused by *A. trifasciata* is unique when compared with the habits of other longhorn beetles that are important pests in Ghana. This beetle actually kills the stem to make it suitable for egg laying. When the stem is girdled, all tissue above the girdle dies and lateral buds begin to develop below the girdle. Repeated girdling of the young growth gives *B. costatum* a very bushy appearance (Plate 49). In the case of *C. pentandra*, which is grown primarily for shade in the north, extensive lateral development may actually increase the shade value. *Analeptes trifasciata* attack on *Eucalyptus*, however, greatly decreases the trees' value by reducing height growth and causing poor form of infested trees. Mortality of *Eucalyptus* can also occur due to *A. trifasciata*.

Pest Management

A simple method for controlling this pest on *B. costatum* is to encourage early burning. Resistant to fire, *B. costatum* burned in October or November, when adult beetle populations are high, results in effective control. *Ceiba pentandra* cannot be treated in this way because of its susceptibility to fire damage. Pest management practices for *A. trifasciata* are dependent on site conditions. Good sanitation may be one method to reduce insect damage. Also vigorously growing plants exude gummy sap which can discourage an adult female from laying eggs, even after an egg niche has been made. One hymenopteran, *Iphiaulax* sp. nr. *melanaria* (Hymenoptera: Braconidae), is parasitic on the larvae of *A. trifasciata* and has been collected at Navrongo (Roberts 1962a, b). The parasitic larvae emerge from the last larval instar of the cerambycid and pupate in individual cocoons within the larval gallery. Between three and five parasitic larvae have been collected from a single host.

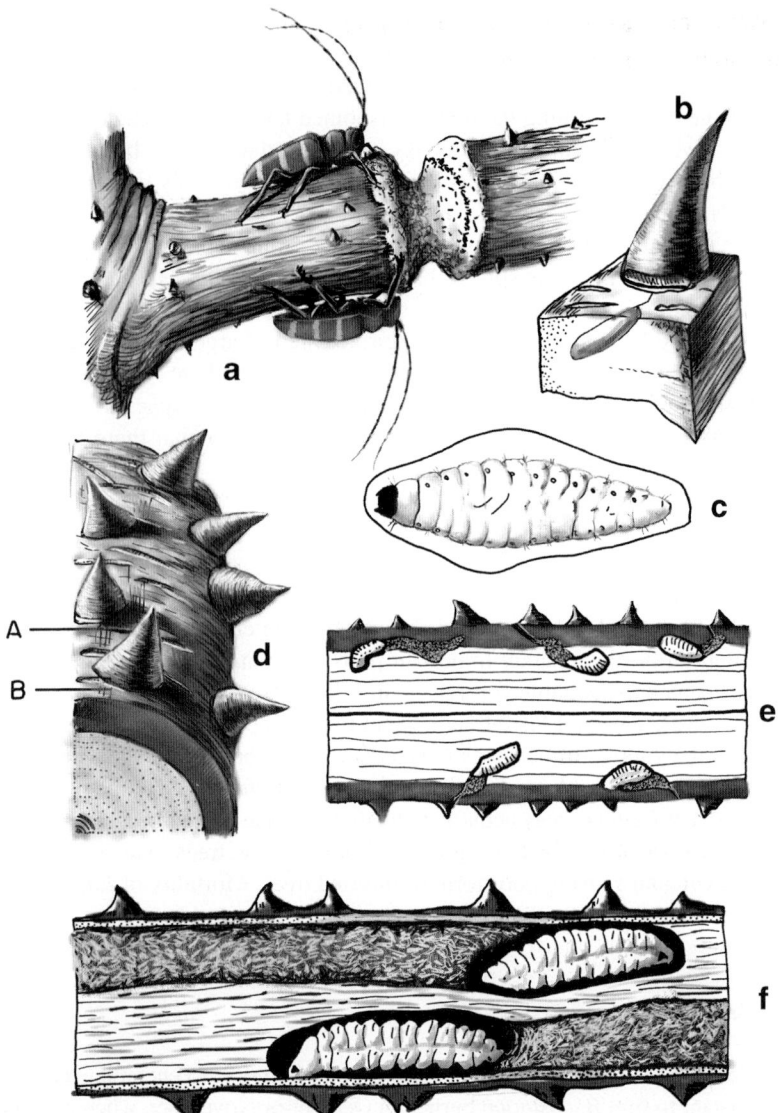

Figure 4.5 Life stages of *Analeptes trifasciata* on branches of *Ceiba pentandra*: (a)egg laid beneath *C. pentandra* spine; (b) egg laid beneath *C. pentandra* spine; (c) first instar larva inside egg capsule; (d) adult feeding scar; (A,B) oviposition scar; (e) young larvae; (f) mature larvae. (From Roberts 1961a, b; reprinted with permission.)

The Borer of Wawa, *Trachyostus ghanaensis*
(Coleoptera: Platypodidae)

Wawa (*T. scleroxylon*) is an extremely important tree species in Ghana. In addition to its wide use locally, it is a major export species and has dominated the export market for many years (TEDB 1987). The most important pest of living wawa trees is *T. ghanaensis* Schedl. *T. ghanaensis* is a member of the ambrosia beetle family Platypodidae. But, unlike ambrosia beetles, which generally attack freshly cut logs, this species attacks living trees (Thompson 1959, 1960).

Description and Life History

The adult beetle is elongate in shape (9 mm long × 3 mm wide) with a dark brown dorsal surface, changing to orange on the ventral surface. The female is slightly larger than the male and the posterior of the elytra (hardened forewings) is not heavily armored like the male. The eggs are white, translucent, ovoid, and approximately 1 mm in size (Roberts 1969). First instar larvae are small, C-shaped, and very active and possess seven pairs of false legs. The last instar is cylindrical in shape and lacks false legs. The last instar is usually the most commonly encountered life stage in any gallery because the larvae pass through the initial instar stages rapidly. The pupa is of the exarate type.

Attack is initiated by the male who selects a host and bores a 2–4 cm long tunnel into the sap wood of the host tree. A few days later the female arrives at the tunnel and mating occurs. The male produces a high-pitched chirping sound by rubbing the dorsal surface of the abdomen on the underside of the elytra. This stridulating appears to attract the female (Thompson 1959).

After mating, the female takes complete responsibility for gallery construction. The male is delegated to the role of frass removal and protection of the tunnel entrance. When an invader attempts to enter the tunnel, the male stridulates loudly, attempting to frighten the intruder, and blocks the tunnel entrance with his armored posterior.

The female continues to extend the gallery for 10–15 cm toward the tree center (Figure 4.6). Then the female turns 90° and the gallery is continued parallel to the circumference of the tree (Figure 4.6). The first cluster of 3–4 eggs is laid at a right angle to the main gallery. These eggs typically hatch and develop to mature larvae before the female continues the tunnel. The first instar larvae bore larval galleries perpendicular to the parent gallery. At the end of these branch tunnels the larvae construct pupal chambers perpendicular to the galleries (Plate 52). There are normally seven instars.

The adult female continues boring the egg gallery and laying eggs periodically. The number of eggs per cluster increases as the tunnel is extended. The first cluster contains 3–4 eggs and the final cluster usually contains 24–30 eggs. Young adults stay within their branch tunnels for a period of time, eventually gathering near the

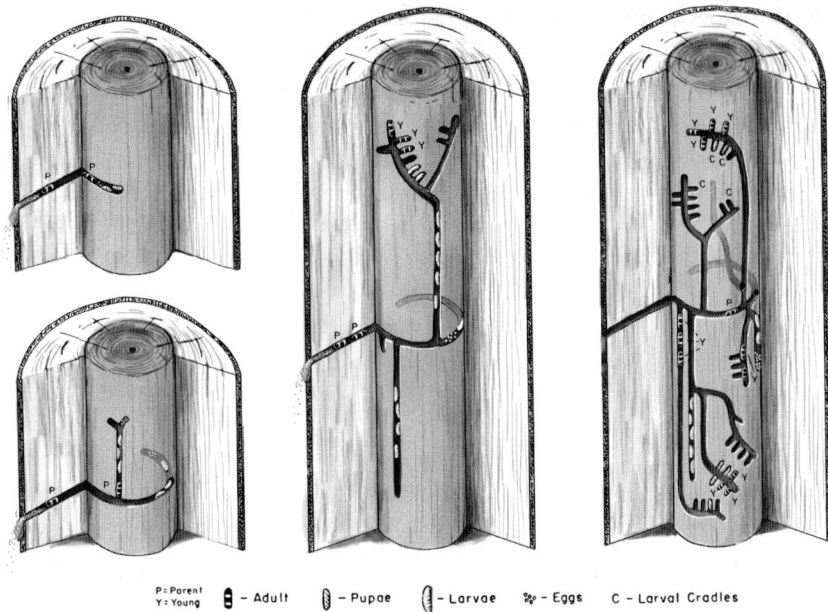

P = Parent   ◐ – Adult   ◖ – Pupae   ◖ – Larvae   ✿ – Eggs   C – Larval Cradles
Y = Young

Figure 4.6 Pattern of gallery formation in wawa by *Trachyostus ghanaensis*

main entrance hole to await suitable weather conditions to fly to new hosts. Adult emergence takes place at intervals over 7–12 months in a single gallery. Evidence now available suggests that the rate of development of *T. ghanaensis*, particularly the larval stages, is much slower than that of other known tropical platypodids. The growth rate of the beetle may be influenced by the slow development of the ambrosia fungus, which in turn may be affected by its living tree habitat (Robert 1961).

T. aterrimus, which attacks unhealthy trees and felled logs, has five larval instars. Initially, the male removes all frass from the tunnel, but as tunnel construction continues the older side tunnels serve as disposal sites for frass. In a completed gallery, all stages of insect development are found and single galleries can be active for up to 24 months.

Damage

The incidence of *T. ghanaensis* attack varies significantly with tree diameter, site conditions, and stand density in Ghana (Robert 1960). Small-diameter trees (30–90 cm) are almost never affected, while trees (1.5–2.0 m) in diameter receive the greatest attack. Wawa, grown on sites with well-drained soils are less often attacked when compared with those growing in low, swampy areas. Stand density also influences attack behavior. In areas of thick underbrush, such as cocoa farms, attack occurs on wawa stems up to the top of the cocoa canopy (5–7 m). Very little attack occurs above that point.

Wawa borer attack normally does not kill the tree or have a major effect on subsequent growth. However, the effect of attack on wood degradation is important. The insect tunnels affect the appearance of the wood and, more importantly, fungi carried into the tree by the adults cause staining and subsequent degradation of lumber. The only external sign of *T. ghanaensis* attack is the adult entrance hole. The hole is approximately 3 mm in diameter and bits of coarse frass are found near it. In lumber, attack can be distinguished by the size of the tunnel and the characteristic tunnel platform.

Records show that the greatest number of attacks occurs before the new flush of leaves appears. At this time, starches that have been stored as food since the last growing season are rapidly being broken down, transported, and serve as attractants for the beetle (Robert 1960).

Pest Management

The general control measure for insects attacking mature trees (rapid extraction and conversion) is impractical here. By law, wawa must exceed 2.0 m in diameter before it can be harvested. The highest incidence of *T. ghanaensis* attack occurs in the 1.5–2.0 m-diameter class.

Silvicultural control is a possible alternative. Choosing well-drained sites and reducing the amount of understory vegetation may reduce incidence of *T. ghanaensis* attack.

Unlike *T. ghanaensis*, a comparatively rich insect fauna has been found in association with the nests of *T. aterrimus*. The following Coleoptera appear to play an important part in natural control; *Macedeum gigianteum* (Krtz.) (predator), *Sosylus validaes* (Krtz.) (pupal parasite), *Bolbocephalus mechowi* Kolbe (nest parasite), and *Scolytroproctus hercules* Mshl. (nest parasite). The nest invaders enter the nest, kill the parent beetles, and establish their own brood within the main gallery (Roberts 1962a, b).

The Ofram Borer, *Doliopygus dubius Samps* (Coleoptera: Platypodidae)

Another valuable tree species, ofram (*T. superba*), is attacked by the ambrosia beetle *Doliopygus dubius*. The distinctive orange-stained wood associated with this beetle's attack is well known. The stain is not from the ambrosia fungus, but appears to be a reaction by the tree to the fungus. *Doliopygus dubius* is found in the northern Ashanti and Brong-Ahafo regions of Ghana. It is also common in certain areas of Nigeria. The ofram borer is a unique ambrosia beetle because it attacks living *T. superba*. The insect also attacks freshly felled logs of many species, but prefers ofram logs.

Description and Life History

Eggs are white, ovoid, and approximately 5.0–5.5 mm in length. The adult male is approximately 5.3–5.5 mm long, 3.6 times as long as it is wide. The elytra are dark brown and appendages are yellowish brown tinged with black. Abdominal sternites are black, smooth, and polished. The female is approximately 6.0 mm long; 3.6 times as long as it is wide. Elytra are a little wider than the pronotum and approximately 1.7 times as long as the pronotum. *Doliopygus dubius* has five larval instars that can be distinguished readily using a key (Browne 1961a, b). The duration of the first four instars is very short; the final instar appears approximately 2 weeks after egg hatch. Newly emerged adults of *D. dubius* fly in the evening from about 7 to 10 p.m. They are strongly attracted to ultraviolet light. The male selects a host tree and bores into the wood. A chamber is excavated to accommodate both the male and a female, who usually joins him immediately. As in other Platypodidae, after mating, the female takes over the role of gallery construction while the male removes the frass. The female bores to a depth of 3–5 cm and then bores along the grain parallel to the stem surface. The female bores an average of 10 mm per day.

When the main gallery is complete, egg laying begins. At weekly intervals the female lays batches of eggs; each successive batch contains a few more eggs than the last batch. Apparently four batches are laid, with the last batch containing at least 50 eggs. Eggs hatch in approximately 3 days and the young larvae immediately begin to feed on the ambrosia fungus. Behavior of the fifth instar larva is very characteristic. Possibly in response to overcrowding, fifth instar larvae begin boring tunnels. Normally, only adult Platypodidae tunnel, but *D. dubius* is an exception. The duration of the fifth stadium is approximately 16 days and includes construction of the pupal cell at the end of the branch tunnel. The pupal stage lasts approximately 10 days. Adults then emerge and seek a new host tree.

Damage

The mode of attack of the ofram borer is considerably different from that of the wawa borer. Generally only a few wawa borers attack a given tree, and at times only one insect per tree is found. However, *D. dubius* will attack in very large numbers, riddling the tree with entrance holes. *T. superba*, however, has the ability to overcome some attacks by exuding large amounts of sap, which kills the insects. This phenomenon is commonly referred to as "pitching out." Even freshly cut logs have the ability to pitch out *D. dubius* for some time after felling. Nevertheless, many ofram trees are successfully attacked by this insect pest.

Infested wood exhibits typical ambrosia beetle damage. Round holes approximately 2.1 mm in diameter are conspicuous with a fungus stain surrounding them. The actual ambrosia fungus leaves a yellowish stain near the edge of the hole, but a grey stain from unrelated staining fungi is also visible (Browne 1961a, b). Approximately 5 mm away from the outside edge of the hole a distinct orange ring appears, presumably the chemical reaction of the tree to insect/fungus attack.

Ofram commonly has a defect called black hearts which consist of discolored wood that is unmarketable. The cause of this defect is unknown, but if it is indeed a fungal attack, the holes of *D. dubius* could serve as the infection court.

Pest Management

As discussed earlier, natural control occurs when the tree pitches out the insects. Genetic selection of trees producing abnormally large quantities of sap is one possible control method. Natural predators of the ofram borer have been studied briefly (Browne 1961a, b). Tailor ants (*Oecophylla* spp.) attack adult beetles as they begin boring into logs. However, these ants forage during the day, while the beetle initiates attack in the early evening. As a result, it is mostly those beetles that entered slowly which are preyed upon. Other predators of *D. dubius* include two beetles in the family Colydiidae, *Sosylus spectabilis* and *Mecedanum sexualis*, and two members in the family Brenthidae, *Diplohoplizes armatus* and *Anisognathus csikii*. Little is known about the effects of these predators on pest population levels. Chemical control measures have not been developed.

Hypothenemus pusillus (Coleoptera: Scolytidae)

A survey of nursery pests shows that this polyphagous beetle is a very common shoot borer in Ghana (Cobbinah 1972a, b). The minute beetle attacks mainly unhealthy seedlings and twigs of *T. grandis, T. ivorensis, C. odorata*, and *Gmelina arborea*.

Description and Life History

The adult *H. pusillus* is a small, elongated, cylindrical beetle ranging from 1.5–2 mm in length. The elytra are approximately 1.5 times as long as their combined width. The head is almost hidden by the prothorax when viewed from above. The head and thoracic shield are black in color while the rest of the body is light brown. The eggs are ovoid, cream to translucent in color, and measure 1–1.5 mm long and approximately 1 mm wide (Jones 1959a, b). The larvae are creamy white in color; C-shaped, thicker in regions of the thoracic and the first few abdominal segments, and pointed at the posterior end. The larvae are legless and have a clearly segmented thorax and a sparse covering of setae. The head capsule is light brown in color and projected downward, with deep brown ocelli. The larvae are approximately 2 mm long and 1 mm wide. Pupae are about the size of mature larvae.

The life cycle is not clearly understood; however, information gathered so far indicates the life cycle is between 3 and 5 weeks in duration. Breeding is continuous, with overlapping generations, so that it is possible to find all life stages at the same time. The eggs are laid in clusters of two to five eggs. The larvae live communally in the tunnel where they pupate in a row separated by small plugs of frass. The young adults usually emerge from the original entrance hole, which is frequently located at a leaf axil and from which a white powdery material is ejected.

Damage

The insect shows no marked host preference. Attack appears to be heavy on seedlings that are weakened by drought, poisoned by chemicals, or have sustained mechanical injury from humans or nature. If attack is heavy, death of the seedling may result. This beetle was associated with over 80% of seedling mortality found in the Apirade nursery (Cobbinah 1972a, b). Attacked seedlings have wilted leaves, brittle shoots, poor root development, and generally impaired growth. The beetles make irregular, longitudinal tunnels usually running the entire length of the shoot. These galleries are a characteristic feature of seedlings attacked by *H. pusillus*.

The insect has been recorded in Ghana, India, Kenya, Malaysia, Nigeria, Sierra Leone, Tanzania, and Uganda. In Malaya, it has been reported on twigs or shoots of seedlings and transplants of *Dryobalandos aromatica, S. macrophylla*, and *T. grandis*. In Ghana, it has been recorded on *Annona squamosa, Antiaris africana, C. pentandra, Mangifera indica, Spondias mombin*, and *Theobroma cacao* (Jones 1959a, b).

Pest Management

Due to the inaccessible habitat of the beetle, external chemical control is not effective. In an experiment conducted at the Apirade nursery, adult beetles were found alive in twigs 2 weeks after brushing attacked seedlings with Aldrex 40. Rapid removal of attacked seedlings appears to be an effective method of reducing attack incidence and spread of attack to other seedlings.

Table 4.2 lists the wood-boring insects of living trees in Ghana. The host plant and known life history information is given for each insect listed.

Table 4.2 Wood-boring insects of living trees in Ghana

Species	Order: family	Host plant	Life history/comments
Acanthophorus spinicornis Fabr.	Coleoptera: Cerambycidae	*Cupressus* spp.	Polyphagous longhorn, usually in dead trees but occasionally attacks unhealthy trees
Acridoschema isidori Chevr.	Coleoptera: Cerambycidae	*Hevea brasilliensis, Albizia zygia, Blighia sapida, Celtis mildbraedii*	
Anaemerus tomentosus Fabr.	Coleoptera: Curculionidae	*Eucalyptus camaldulensis E. torreliana*	Mainly found in Guinea Savannah Zone
Analeptes trifasciata Fabr.	Coleoptera: Bostrychidae	*Ceiba pentandra, Tectona grandis, Bombax costatum, Spondias morbin, Eucalyptus alba, E. tereticornis, Lannea nigritana, Lannea kerstingii, Adansonia digitata, Anacardium occidentale*	Adult feeds by gnawing the bark and the underlying wood. When the stem is girdled all tissues above the girdle die. Control: good sanitation. For fire resistant *B. costatum*, burning is an effective control measures
Apate monachus Fabr.	Coleoptera: Bostrychidae	*Dalbergia sisoo, Azadirachta indica, Antiaris Africana, Terminalia ivorensis, Sesbania grandiflora, Baphia nitida, Samanea saman, Glyricidia maculate, Cassia nodasa*	Attacks young and vigorously growing trees; life cycle duration of 5 months recorded but may be longer under adverse conditions (Roberts 1969); control methods including sanitation of plantation sites
A. terebrans Pallas	Coleopteran: Bostrychidae	*T. ivorensis, T.grandis, Cedrela odorata, Triplochiton scleroxylon, Eucalyptus polycarpa, Melia composita, Khaya senegalensis, Poinciana regia*	
Azygophleps scalaris (synonym: *Phragmatoeria scalaris*)	Lepidoptera: Cossidae	*S. grandiflora*	
Bostrychoplites cylindricus Fhs.	Coleoptera: Bostrychidae	*Cassia* spp.	Normally attacks weakened stems; point of entry about 30 cm from the ground
B. productus Imhoff	Coleoptera: Bostrychidae	*Baphia pubescens, Bosquea angolensis*	Same as above
Bostrychopsis tonsa Imhoff	Coleoptera: Bostrychidae	*T. ivorensis*	Usually breeds in dead wood but also tunnels in small living trees, rendering them susceptible to breakage

(continued)

Table 4.2 (continued)

Species	Order: family	Host plant	Life history/comments
Chaetastus tuberculatus Chapius (synonym: *Symmerus tuberculatus*)	Coleoptera: Platypodidae	*Albizzia zygia, Eucalyptus* spp., *T. ivorensis, Khaya grandifoliola*	Usually breed in logs, but sometimes unhealthy trees are attacked as well
Cledus obesus Hustarehe (synonym: *Osphilia obesa*)	Coleoptera: Curculionidae	*K. grandifoliola, K. ivorensis*	A grey or yellowish weevil with brown or black spots; the larvae tunnel in the stem branch axils, causing stem to swell at point of infestation
Cordylomera spinicornis Fabr.	Coleoptera: Cerambycidae	*K. grandifoliola*	Normally breeds in logs but may infest mature standing trees as well
Curimosphera senegalensis (Haag)	Coleoptera: Tenebrionidae	*Butyrospermun paradoxum*	
Diapus guinguespinatus Chapius	Coleoptera: Cerambycidae	*Unknown*	Polyphagous beetle; attacks mainly dying trees, showing little selectivity among potential food plants
Doliopygus conradti	Coleoptera: Platypodidae	*C. odorata, Eucalyptus angolense*	
D. dubius (Samps)	Coleoptera: Platypodidae	*T. superba*	Commonly referred to as the ofram borer; life cycle is completed between 6–7 weeks; attacked wood exhibits the typical ambrosia beetle damage; similar type of damage has been recorded for the other species of *Doliopygus*
D. erichzoni Chapius	Coleoptera: Platypodidae	*T. grandis*	
D. serratus Strohmeyer	Coleoptera: Platypodidae	*Sterculia rhinopetala*	
D. solidus Schedl	Coleoptera: Platypodidae	*S. rhinopetala*	
D. unispinosus Schedl	Coleoptera: Platypodidae	*A. indica, T. scleroxylon*	
Eulophonotus obesus (Karsh)	Lepidoptera: Cossidae	*T. scleroxylon*	
E. myrmyleon Fldr.	Lepidoptera: Cossidae	*Cola nitida*	
Gyroptera robertsi Bradley	Lepidoptera: Pyralidae	*Entandrophragma cylindricum, K. grandifoliola, K. ivorensis, Swietenia macrophylla*	Life cycle takes 2–3 months; multivoltine; unlike *Hypsipyla robusta*, this pyralid does not attack green shoots; attacks mainly old trees

Species	Order: Family	Host	Notes
Hypothenemus eroditus Westwood *H. pussilus* Westwood	Coleoptera: Scolytidae	*T. grandis, A. africana, C. pentandra, T. ivorensis, Cedrela and Gmelina, Annona squamosa*	A minute beetle which attacks mainly unhealthy seedlings and twigs; life cycle completed in 3–5 weeks; multivoltine, control: removal of infested seedlings; Aldrex 40 has limited success
Hypsipyla robusta (Moore)	Lepidoptera: Pyralidae	*Khaya anthotheca, K. grandifoliola, K. ivorensis, K. senegalensis, Entandrophragma angolense, E. cylindricum*	Known in West Africa as mahogany shoot borer; four instars have been recorded and development from egg to adult is 4–6 weeks; several parasites have been recorded but the two most important are the braconid *Macrocentrius* sp. and *Protomicroplitis* sp.; chemical control: brushing Bidrin on affected parts
Mallodon downesi Hope	Coleoptera: Cerambycidae	*Hevea brasiliensis*	
Monachamus antralis Duvivier	Coleoptera: Cerambycidae	*Petersianthus macrocarpus* (synonym *Combretodendron* spp.)	
M. ruspator Fabr.	Coleoptera: Cerambycidae	*Cupressus macrocarpa*	Life cycle duration is about 4–5 months; adult girdles side branches resulting in the death of tissue above the girdle
M. scabiosus Quedenfeldt	Coleoptera: Cerambycidae	*P. macrocarpus* (synonym *Combretodendron* spp.)	
Orygophora mediofoveata Hampson	Lepidoptera: Noctuidae	*Nauclea diderrichii*	Life cycle takes 3–4 months; larvae favor young shoots; natural enemies include *Macrocentrius* sp. and *Pristomerus* sp.
Periomatus sp. *Chapius*	Coleoptera: Platypodidae		Multivoltine; normally breeds in fallen or unhealthy trees, some infest healthy trees during periods of stress; adults are nocturnal and short-lived; kills host through repeated infestation
Petronagha gigas	Coleoptera: Cerambycidae	*Casuarina* sp., *C. pentandra*	
Phryneta leprosa Fabr.	Coleoptera: Cerambycidae	Moraceae (*Morus alba*)	A brown longhorn; normally breeds in dead fallen trees but sometimes attacks healthy trees

(continued)

Table 4.2 (continued)

Species	Order: family	Host plant	Life history/comments
Platypus hintzi (Schaufuss)	Coleoptera: Platypodidae	*Casuarina* sp., *Eucalyptus* sp.	Life cycle is completed within 5–7 weeks; multivoltine; normally infests recently fallen trees but attack on healthy trees may occur during periods of temporarily decreased vigor, especially during the dry season
P. refertus Schedl	Coleoptera: Platypodidae	*Casuarina* sp.	
Polygraphus granulatus Egg	Coleoptera: Scolytidae	*Cedrela mexicana*	Bark beetle that can cause serious injury to weakened trees
Terastria reticulata Gn.	Lepidoptera: Pyralidae	*Erythrina addisoniae*	Life cycle takes 6–7 weeks; multivoltine; attacked plants usually have multiple leaders; 1% Bidrin gave 100% control
Trachyostus aterrimus Schaufuss	Coleoptera: Platypodidae	*T. superba, Cola gigantean, Afzelia africana, C. mildbraedii, C. zankeri, C. mildbraedii, Albizia fermginea*	
T. carinatus	Coleoptera: Platypodidae	Same as above	
T. ghanaensis	Coleoptera: Platypodidae	Same as above	Common name: wawa borer; cause severe degrade of wawa by the presence of tunnels and associated fungal stains
T. tomentosus	Coleoptera: Platypodidae	Same as above	
Tragocephala gorilla Thoms	Coleoptera: Cerambycidae	*Acalypha* spp.	
T. nobilis Fabr.	Coleoptera: Cerambycidae	*Acalypha* spp.	A black and yellow longhorn beetle; adults attack branches and coppice shoots often girdling them; life cycle takes about 2 years; normally breeds in decaying logs, however, severe attack on young trees sometimes occurs

Species	Order: Family	Host	Notes
Tridesmodes ramiculata Warren	Lepidoptera: Thyrididae	*T. ivorensis*	Common name – emire shoot borer; life cycle is completed in 1–3 months; natural enemies include lizards and *Mydaea superba* Stein; chemical control: Bidrin brushed on affected parts effective
Xyloborus compactus (Synonym: *Xylosandrus compactus*)	Coleoptera: Scolytidae	*Aucoumea klaineana, Khaya ivorensis*	Small ambrosia beetle; usually breeds in newly cut logs of dying trees but also infest temporarily stressed or injured trees; attacks mainly transplants that have yet to recover from transplanting shock
Xyloborus mortatti Haag	Coleoptera: Scolytidae	*Alchornia cordifolia, Cannia* sp.	
Xyloborus perforans Wollaston (Synonym *X. testaceus*)	Coleoptera: Scolytidae	*K. ivorensis Entandrophagma utile*	Small ambrosia beetle; usually breeds in newly cut logs or dying trees; but also infest temporarily stressed or injured trees; attacks mainly transplants that have yet to recover from transplanting shock
X. semiopacus Eichhoff	Coleoptera: Scolytidae	*K. ivorensis*	Same as above
X. sharpae	Coleoptera: Scolytidae	*K. ivorensis*	Same as above
Xyloctonus quadricinctus Schedl	Coleoptera: Scolytidae	*Butyrospermum parkii* (Shea butter tree), *Chrysophyllum albidum*	Bark beetle; larvae usually tunnel in the direction of the wood grain
X. scolytoides Eichhoff	Coleoptera: Scolytidae	Same as above	Same as above
Xylopertha crinitarsis Imh.	Coleoptera: Bostrychidae	*Gliricidia sepium, A. africana, Milicia excelsa, Triplichiton scleroxylon*	Development from egg to adult takes approximately 3 months; this group has been recorded on many forest tree species; normally attacks felled trees
X. picea Ol	Coleoptera: Bostrychidae	*B. nitida*	
Xyloperthodes orthogonius Lesne	Coleoptera: Bostrychidae	*Albizia* spp., *Cassia* sp., *T. scleroxylon*	
X. nitidipennis Murr	Coleoptera: Bostrychidae	*Bauhinia tomentosa, Albizia* spp.	Life cycle takes approximately 3 months; normally attacks recently felled trees or living trees weakened or damaged by fire

Chapter 5
Pests of Flowers, Fruits, and Seeds

Introduction

In the natural forest, pests of reproductive structures (flowers, fruits, and seeds) tend to be of little economic importance. Most tree species produce far more seed than is necessary to establish regeneration. Other factors such as competition tend to be far more important in tropical forests. However, when high value species are artificially regenerated through the production and planting of seedlings, a reliable seed source becomes critical. The need to improve the seed production capability is well established in West Africa (Okoro and Dada 1987; Ouedraogo and Verwey 1987). Many species like wawa are known to have irregular seed years (Taylor 1960; Danso 1970; Kudler and Jones 1970). Insects that infest seeds can be a significant factor in this variability in seed production (Kudler and Jones 1970). The significant role insects play in reducing the availability of seeds is recognized in other parts of Africa as well (Shakacite 1987; Hassani and Messaoudi 1986; Ross 1979; Rasplus 1988).

The natural variation in seed production can be affected by insect pests. When seed production is low, the pest population is also low because of a limited food resource. The first heavy seed crop following a low crop period is usually relatively free from damage because the insect population has not yet increased in response to the increased food supply. However, if good seed crops continue for a few seasons, insect populations can build to the point where 100% of the seed crop is killed. Under this latter set of conditions, it is possible to have years when no seed is available. This is especially serious because many species of Ghanaian trees have seeds that do not store well (Gyimah 1984). Logistics of plantation establishment can be greatly affected by uncertain seed supplies.

A second situation where pests of reproductive structures become severe is in seed orchards. Researchers at the Forestry Research Institute of Ghana have established clonal seed orchards of selected superior teak. *Tectonia grandis*. In this situation where relatively few individual trees may ultimately provide a high proportion of the seed for regeneration, even small amounts of damage become intolerable. As tree improvement practices improve in Ghana, seed pests will become increasingly important.

Seed pests can also become a problem during storage. There are many factors to consider in developing a suitable seed storage system, including insects (Jones and Damptey 1969). The three insect species groups discussed in this chapter (guarea fruit weevil, *Menechamus* sp.; wawa fruit borer, *Apion ghanaensis* and *A. nithonomiodes*; emire fruit weevil, *Nanophyes* sp.) are of major importance because they attack both reproductive structures in the field and seeds in storage. Without proper management of insects species that attack stored seeds, large quantities of valuable seed can be completely destroyed.

Exotic tree species can occasionally be attacked by native reproductive pests. *Senna cassia siamea* is a potentially valuable fuel wood, shelter wood, and agroforestry species introduced to Ghana (Nkansa-Kyere 1972). The reproductive structures of this species have been attacked by *Mussidia nigrivenella* Ragonot and *Thylacoptile paurosema* (Meyrick) and by *Eublemma* sp., a Noctuidae. *Eublemma* sp. likewise attacks the flowers and fruits of *Eucalyptus* sp., in Ghana. Most likely these are native pests that have adapted to the new hosts. *Selepa docilis* is an interesting Noctuidae that is a common agricultural pest of garden egg (eggplant) in Ghana that has been found damaging wawa flowers. As can be seen from Table 5.1 of this chapter, there are a multitude of insect species that can attack and damage seeds of forest trees. For example, wawa alone has 11 species of insects that attack its reproductive structures. The list presented in Table 5.1 probably understates the diversity of species because most forest tree species probably have several insects attacking the nutrient-rich reproductive structures.

A phenomenon that is common throughout the world is that only when countries begin serious efforts at artificial regeneration does the importance of seed-infesting pests become obvious. As Ghana expands its plantation projects, many more species of insects attacking reproductive structures will be discovered. Control of these pests is most commonly achieved by cultural and chemical methods. Early collection of fruits may be effective methods of reducing *Apion* spp. damage to wawa (Danso 1970). Also isolating seed-producing areas from natural stands may reduce the immigration of pests into a seed production area. When high-value seeds are at stake, chemical methods may be justified. Effective chemical control requires a thorough knowledge of the pest complex, their respective life histories, and damage potential. There can be little doubt that when genetically improved tree seeds are involved, pest control practices are easily justified. Some of the better-studied fruit and seed pests occurring in Ghana are discussed below.

Guarea Fruit Weevil, *Menechmaus sp.* (Coleoptera: Curculionidae)

Guarea cedrata is one of several trees which have been included in species trial programs in Ghana. Because efforts are being made to regenerate *G. cedrata* in nurseries, attention has been drawn to seed viability. The curculionid *Menechamus* sp., in addition to other animals, has been found feeding in *G. cedrata* fruits.

Description and Life History

This weevil was identified as *Menechamus* sp. and bears a close resemblance to *M. discrepans* Faust, which is found in Gabon (Kudler 1970a–c). The adults are 3–4 mm long with a curved snout about one third the length of the body (Figure 5.1). The body is entirely covered with slight brownish pubescence, some of which is usually lost on old individuals. The visible parts of the body are dark brown in color. Antennae are clavate, attached laterally to the mid-point of the beak, and elbowed, with the basal segments resting in a groove along each side of the beak. The pronotum is strongly narrowed toward the anterior, with a dorsal line slightly raised. The elytra are wider and longer than the pronotum. There are ten distinctive longitudinal striae on each elytron. The femur possesses a larger spur on the inner surface and the tibia has a spine at the distal end.

Eggs are oval, whitish, and less than 0.5 mm long. Larvae are yellowish, somewhat curved, and legless (Figure 5.1). The heads are slightly brown, with dark brown mouthparts. Mature larvae are 3–4 mm long, with a row of single hairs on each segment. Pupae resemble the adults, but are creamy white and have wing pads in place of elytra (Figure 5.1).

Kudler (1970a–c) described the life history of *Menechamus* sp. The insect requires 4 weeks to complete its life cycle, and damage is the result of several overlapping generations. Newly emerged adults stay in their pupal cradles for approximately 2–3 days to mature, feeding on the pulpy fruit tissue. After emergence, the adults seek mates. The females puncture the epidermis of the fruit with their snout and lay eggs. The larvae hatch in 1–2 days and feed on the fleshy tissue. When the larvae mature, they construct cradles stuffed with excrement within which they pupate.

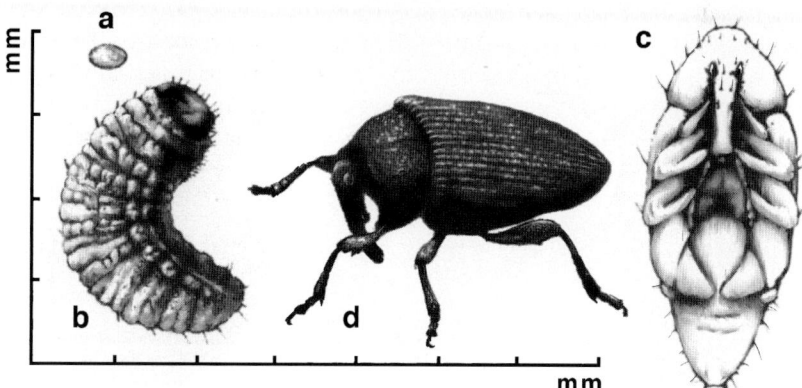

Figure 5.1 Guarea fruit weevil, *Menechamus* sp.: (a) egg; (b) larva; (c) pupa; (d) adult. (From Kudler 1970a–c; reprinted with permission from FPRI, Kumasi.)

Damage

Because of the short life cycle, the weevil attack, once started, increases very rapidly. Fruits that were 25% attacked in November were 100% infested by January (Kudler 1970a–c). Heavy infestation results in the entire fruit being consumed, leaving only a mass of excrement. Even a few weevil holes can result in damage by allowing the entrance of fungi. Damage can continue in stored fruit, but new attacks do not occur on fully dried seeds (Figure 5.2). In seed orchards, early selection appears to be the best method of insuring a high percentage of sound fruits. Only mature, uninfected seeds should be collected and stored to avoid contamination from weevils. Chemical fumigants can kill residual insects in seeds and will protect the remaining seed until it is fully dried.

Fruit Borer on Wawa, *Apion ghanaensis* and *A. nithonomiodes* (Coleoptera: Apionidae)

Wawa (*Triplochiton scleroxylon*) is one of the most valuable tree species in Ghana. Its wood is widely used as concrete forms and in general construction. In spite of

Figure 5.2 Dried seeds of *Guarea cedrata*. Both fresh and dried seeds are susceptible to attack from *Menechamus* sp.

Figure 5.3 *Triplochiton scleroxylon* fruit seeds with holes due to attack by the fruit borer of wawa

its high value, much difficulty has been encountered in establishing wawa planta tions. One of several major factors is the availability of viable seed. Taylor (1960) reported that wawa fruiting was irregular, and sound seed was difficult to obtain. Later Kudler and Jones (1970) discussed the presence of weevils in wawa seeds (Figure 5.3). The closely related weevil, *Apion (Pseudopion) ghanaensis* and *A. nithonomiodes* (Voss), are responsible for the damage. *Apion ghanaensis* is the more abundant species and is described by Kudler and Jones (1970).

Description and Life History

The adults are blackish in color and covered with sparse, light brown pubescence (Figure 5.4). Body size and snout length distinguish between males and females. The body of the female is approximately 3 mm long with a slightly curved snout approxi- mately 1.5 mm long. Males are thinner and smaller, being slightly more than 2 mm long with snouts about half the body length. Eggs are oval, semitransparent, and less

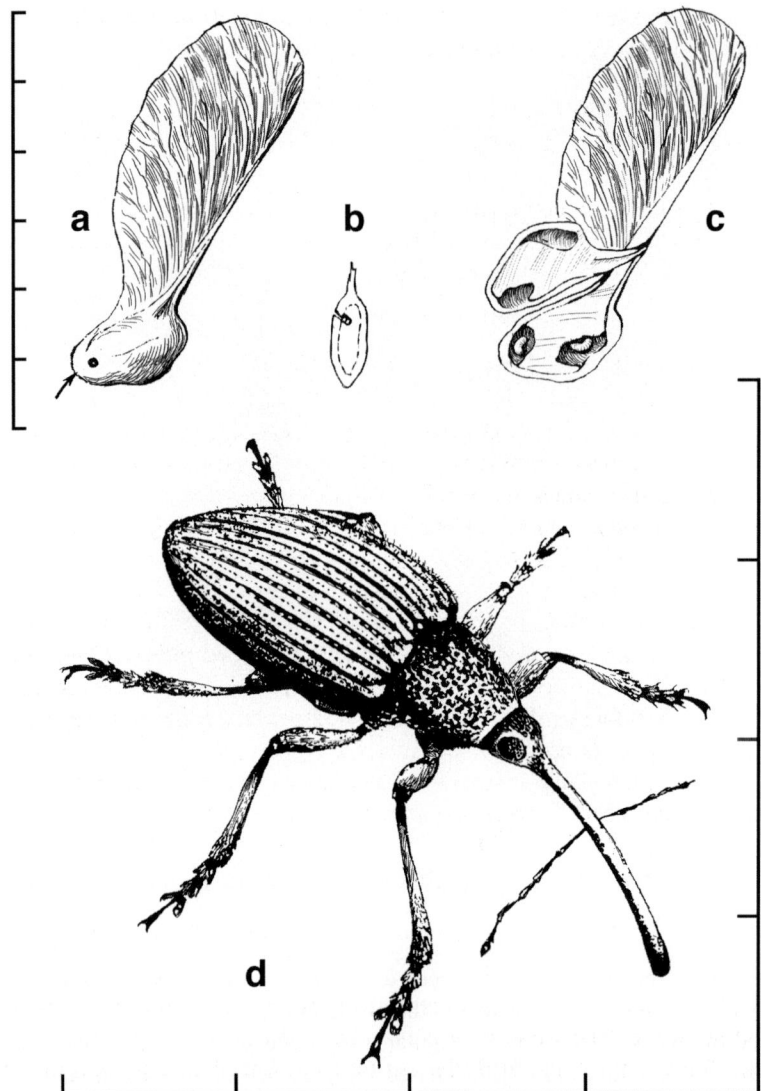

Figure 5.4 Adult weevil (d), exit hole (a), puncture through the fruit (b), and larval feeding (c) of the fruit borer of wawa. (From Kudler and Jones 1971; reprinted with permission of FPRI, Kumasi.)

than 0.5 mm long. Larvae are yellowish white, wrinkled, fat, and legless; mature larvae are 2.5–3.0 mm long. Pupae are creamy white and resemble the adults.

The adults emerge during the fruiting period of wawa. Females puncture holes in the epidermis of fruits with their snouts and eggs are deposited in the holes. After 1–2 days the larvae hatch and begin feeding on the seed tissue. Several grubs may infest the same fruit. Pupation occurs inside the fruit. Mature adults make an emergence hole with their snout. The entire life cycle takes 20–30 days and three or four generations may occur in the normal fruiting period of wawa.

Damage

Blackish spots, excrement, or exit holes in the epidermis of fruit indicate insect attack (Figure 5.3). Feeding by *A. ghanaensis* and *A. nithonomiodes* on the seed tissues destroy potentially valuable seed.

Pest Management

There are some natural factors that have a controlling effect on the weevil population. Starvation often occurs when several grubs are present in the same seed. A few parasites have been found to attack the weevil, including a eupelmid, *Eupelmus* sp.; two eulophids, *Entedon apionidis* Ferr and *Tetrastichus* sp.; two braconids, *Bracon* sp. and *Triaspsis* sp.; and a pteromalid (Kudler and Jones 1970). Although parasites kill many weevils, they are not effective biocontrol agents. A silvicultural method of control suggested by Danso (1970) is to carefully watch the development of seeds on the tree and collect the seeds just as they reach full development. Felling the tree, then collecting and artificially drying the fruits have resulted in 69% germination. This method could be used if a suitable method of early collection is employed so that felling of trees is not required. Kudler and Jones (1970) recognized the very brief period of seed viability in wawa. Consequently, careful determination of seed maturation and immediate fruit harvesting will reduce weevil damage.

Emire Seed Weevil, *Nanophyes sp.* (Coleoptera: Curculionidae)

Emire, *Terminalia ivorensis*, is an important forest species that has been widely planted in Ghana. The wood is of high quality and is widely accepted by woodworkers. Taylor (1960) indicated that the fruits are abundant, but seed germination is poor. Often large numbers of immature fruits can be found beneath trees and many of these have insect damage (Jones 1969) (Figure 5.5). Though germination of emire can be improved by cultural practices (Jones 1969), insects still significantly affect the availability of seed for this species. The most important insect pest is the emire seed weevil.

Description and Life History

The weevil is a curculionid belonging to the subfamily Nanophyinae. It has been identified as *Nanophyes* sp. and is closely related to *N. inturiensis*, a curculionid

Figure 5.5 Emire, *Terminalia ivorensis*, seeds with evidence of damage from the emire seed weevil, *Nanophyes* sp.

occurring in the Central African Republic (Jones and Kudler 1969). Jones and Kudler (1968) described the weevil as slightly less than 3 mm long with a curved snout approximately half the body length (Figure 5.6). It is entirely brown, pubescent and each elytron is marked with seven distinct longitudinal black mottling. Four spurs on the inner surface of the front femur are a distinctive characteristic. The larvae are orange-yellow, wrinkled, fat, and legless. Mature larvae attain a length of 2.5–3.0 mm. Eggs are deposited while the seeds are still on the tree. As is typical for weevils, a hole is made with the snout through the seed coat, within which the eggs are laid. The larvae feed on the young seed for a period of 26–31 days (Figure 5.6). Pupation occurs in the seed and lasts for 6–8 days. The entire life cycle may be completed in 6 weeks.

Damage

A dark spot approximately 1 mm in diameter, packed with excrement, is observed on attacked seeds. Seeds have 1–7 holes, normally at the part of the seed furthest from the attachment point. Fresh holes have drops of pale yellowish sap that become darker with age.

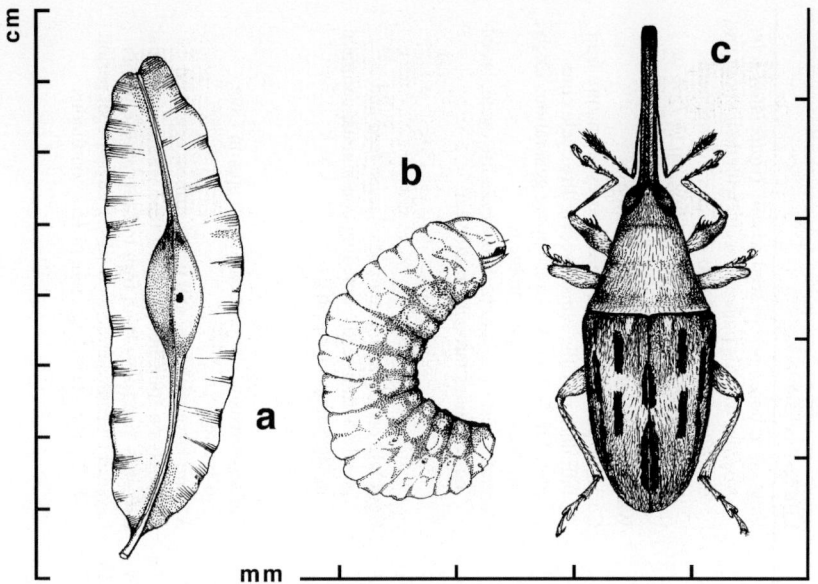

Figure 5.6 Damage to emire seed (a), mature larva (b), and adult (c) of the emire seed weevil *Nanophyes* sp. (From Jones and Kudler 1968; reprinted with permission of FPRI, Kumasi.)

Extensive damage can occur to the seeds of *T. ivorensis*. At Mpraeso, a collection of seeds was found to be only 9% sound due to attack by this weevil (Jones 1969). Germination percentages of 40% and below are common because of the emire seed weevil. Additional damage occurs when fruits fall prematurely due to weevil attack.

Pest Management

As yet no control measure has been developed for *Nanophyes* sp. Needless to say, if emire is intended for large-scale plantings, more work is needed. These and other insect pest of flowers, fruits, and seeds in Ghana are listed in Table 5.1. Host plants and known life cycle information are given for each insect species listed.

Table 5.1 Insect pests of flowers, fruits, and seeds in Ghanaian forest

Species	Order: family	Host plant	Life history/comments
Apion ghanaensis Voss	Coleoptera: Apionidae	Triplochiton scleroxylon	Wawa fruit borer, severe seed pest of Triplochiton scleroxylon; weevils puncture seeds with their snout and deposit eggs; life cycle 20–30 days; four generations during T. scleroxylon fruiting; up to 75% of seed damaged
Apion nithonomoides Voss	Coleoptera: Apionidae	T. scleroxylon	Wawa fruit borer, same as A. ghanaensis
Auletobius kentzeni M.	Coleoptera: Curculionidae	Terminalia ivorensis	Common pest of fruits on trees; causes premature fruit fall; dark metallic bluish weevil; life cycle completed in 5–7 weeks; overlapping generations likely; can be a serious pest problem
Balanogastris kolas (Desh.)	Coleoptera: Curculionidae	Guarea cedrata, Cola spp.	Common pest of cola nuts; 5–6 weeks life cycle; breeding continues throughout the year
Baris sp.		G. cedrata	
Bocchoris inspersalis Zell.	Lepidoptera: Pyralidae	T. scleroxylon	
Bruchidius uberatus (Fahraeus)	Coleoptera: Bruchidae	Acacia nilotica	Eggs laid on dry and green pods; larvae feed and pupate within seeds; breeding continuous; common in the Sudan savannah
Calama sp.		T. scleroxylon	
Caryedon albonotatum Pic			
Caryedon cassiae (Gyll.)	Coleoptera: Bruchidae	Prosopis africana, Cassia spp.	Causes extensive damage on Cassia spp. in Côte d'Ivoire
Catopyla dysorphnaea Bradly	Lepidoptera: Pyralidae	Entandropiragma angolense, Entandrophragma candollei Entandrophragma cylindricum Entandrophragma utile, Khaya grandifoliola, Khaya ivorensis, Khaya senegalensis, Lovoa trichilioides, Pseudocedrela kotschyi	Mature larva with blue and white stripes; eggs laid singly on fruit; larvae at first feed within individual seeds then move from fruit to fruit; pupate within seed on fruit; larvae feed 14–21 days; pupal period 8–22 days; breeding more or less continuous

Species	Order: Family	Host	Remarks
Characoma nilotica Hamps.	Lepidoptera: Noctuidae	*A. nilotica, Cola caricifolia, T. scleroxylon*	Green fruits attacked; apparently short life cycle
Clyphipterix sp.	Lepidoptera: Cosmopterigidae	*T. scleroxylon*	Small insect that mines leaves and bores in stems and seeds; occasionally preys on scale insects
Cryptoblabes gnidiella (Mill.)	Lepidoptera: Pyralidae	*K. senegalensis*	1 cm red larva feeds within green and dried fruit; 4–11 white compact cocoons within a single fruit; adults emerge during January–June; have been reared from *Phytolyma* galls on *Milicia excelsa*
Crytophlebia leucotreta Meyr.	Lepidoptera: Tortricidae	*Parkia clappertoniana*	Widely distributed moth in Africa south of Sahara; brown and white moth; adults nocturnal; each adult produces approximately 100 eggs; conspicuous frass; pupation in cocoons in the soil; pest of citrus and stone fruit
Crytophlebia peltastica Meyr.	Lepidoptera: Tortricidae	*P. clappertoniana*	Has been observed boring into inflorescence (fruit) of its host
Diamerus impar Chap.	Coleoptera: Scolytidae	*Ficus* sp.	
Diclidophloebia eastopi	Homoptera: Psyllidae	*T. scleroxylon*	See leaf-feeding section for biological details; this insect weakens the stem bearing seeds
Earias biplaga Walker	Lepidoptera: Noctuidae	*T. scleroxylon*	
Earias ogovana Holl.	Lepidoptera: Noctuidae	*T. scleroxylon*	
Eublemma sp.	Lepidoptera: Noctuidae	*Cassia siamea, Eucalyptus* sp., *T. scleroxylon*	Larvae feed on inflorescences and fruits; larvae common during November–February; adults emerge during December–April; larval stage 12 days; pupal stage 8–12 days; mature caterpillar 4–5 cm, pale yellow with white lines
Hypothenemus polyphagus (Egg)	Coleoptera: Scolytidae		This bark beetle will infest almost any leathery or woody fruit
Hypothenemus uniseriatus (Egg)	Coleoptera: Scolytidae	*Cassia siamea*	Seed pest in other parts of Africa and likely in Ghana; currently known as twig borer in Ghana

(continued)

Table 5.1 (continued)

Species	Order: family	Host plant	Life history/comments
Hypsipyla robusta (Moore)	Lepidoptera: Pyralidae	*Carapa grandiflora, Carapa procera, E. cylindricum, K. grandifoliola, K. ivorensis, K. senegalensis*	Species also feeds in shoots; eggs laid on green seeds and larvae bore through seeds; mature larvae move from fruit to fruit; attack hole surrounded by frass, gum, and silk; pupation in white cocoons within the fruit; one generation/year; larval development longer in fruits than shoot
Laspeyresia sp. nr. *tricentra* Meyr.	Lepidoptera: Tortricidae	*Pericopsis elata*	Observed feeding in seeds
Lophocrama phoenicochlora Hamps	Lepidoptera: Noctuidae	*T. scleroxylon*	
Menechamus sp. n. *discrepans* Faust	Coleoptera: Curculionidae	*G. cedrata, Cola gigantea*	Guarea fruit weevil, serious pest that can infest 100% of seeds; life cycle 4 weeks; freshly hatched adults remain in cradles for 2–3 days to mature; mating occurs on the fruit; eggs are laid in feeding punctures
Mussidia nigrivenella Ragenot	Lepidoptera: Pyralidae	*Amblygonocarpus andongensis, C. siamea, C. sieberiana, Diospyros mespiliformis, Entandrophragma* spp., *Gardenia imperialis, P. clappertoniana, P. africana, Tetrapleura tetraptera*	Common moth; breeds in green and dried fruits; larvae feed on seed and fruit wall; cocoon within fruit; breeds continuously
Nanophyes sp. n. *ituriensis* Hust	Coleopteran: Curculionidae	*T. ivorensis*	Emire seed weevil; typical weevil that attacks seeds on the trees; eggs laid in feeding puncture; light brown weevil; 6-week life cycle; up to 50% of seeds damaged; causes premature fruit drop; punctures penetrate the wall of the fruit case and are concentrated at the position of the developing seed; life cycle completed in 3 months
Poecilips advena	Coleoptera: Scolytidae		

Species	Order: Family	Host	Notes
Poecilips asper	Coleoptera: Scolytidae		October adult flight throughout forest zone
Poecilips crassiventris	Coleoptera: Scolytidae	Unknown	This species flies in January and February; hosts are unknown but the insect is probably a seed pest like others in the genus
Poecilips sannio Schauf	Coleoptera: Scolytidae	*Afzelia africana, Afzelia bella, C. procera, K. ivorensis, Parkia bicolor, Sterculia oblonga*	Female makes irregular winding gallery in which eggs are laid; common species; adults emerge from entrance hole; also breeds in the bark of felled trees; will attack almost any large fruit or seed when dry; does not normally attack fruit on the tree
Poecilips sierrateonensis	Coleoptera: Scolytidae	Unknown	This species flies in May; host unknown but insect is probably a seed pest like others in the genus
Poecilips crassiventris	Coleoptera: Scolytidae	Unknown	
Selepa docilis	Lepidoptera: Noctuidae	*T. scleroxylon*	A common agricultural pest of eggplant in Ghana that occasionally also attacks *T. scleroxylon* seeds; caterpillars can kill a wawa flower bud flower bud at the rate of one a day
Specularius impressithorax Pic	Coleoptera: Bruchidae	*Erythrina senegalensis*	
Thylacoptile paurosema	Lepidoptera: Pyralidae	*C. siamea*	Larvae feed on flowers and fruit; mature larvae 3–4 cm; pale green with double transverse stripe; head light green; larvae feed in groups
Torrtrix dinota Meyr.	Lepidoptera: Tortricidae	*T. scleroxylon*	Larva of this moth feeds on foliage of numerous plants; webs leaves together; defoliates *T. ivorensis*; larvae feed on flowers and fruits of *T. scleroxylon*
Trachylepidia sp.	Lepidoptera: Pyralidae	*P. africana*	
Xyleborus affinis Eichh.	Coleoptera: Scolytidae		Common observed boring into large woody or leathery fruits
Xyleborus alluaudi Schauf.	Coleoptera: Scolytidae	Very broad host range	Common in large, hard fruits

Chapter 6
Pests of Logs, Lumber, and Forest Products

Introduction

In Chapter 5, we discussed the ecological group of organisms that bore into the woody tissue of living trees. In this section we will discuss insects that generally attack wood after trees have been cut or processed into wood products. This arbitrary ecological categorization does not necessarily follow taxonomic lines. For example, in the previous section, we discussed insects representing the families Scolytidae, Platypodidae, Cerambycidae, and Bostrichidae that damage living trees. We will again discuss members of these families that have feeding habits quite distinct from their taxonomic relatives. Most of the examples cited in the previous section that are from these families are exceptions to the general ecological pattern for these taxonomic groups.

Insects associated with wood and wood products are economically important because once the trees are felled they increase in value. An isolated *Khaya ivorensis* tree in the forest has relatively low value in comparison to a truck load of finished furniture made from that species. Also, insects that attack logs result in a significant reduction in the useable sawmill output. Sawmill conversion ratios of 50% are common and a significant part of the half that is lost can be attributed to this group of insects (Kudler 1967). The problem becomes particularly apparent when entire shiploads of logs or lumber are rejected at their destination because of insect infestation. There can be little doubt that wood-boring insects are responsible for tremendous economic losses to forestry throughout the world.

Termites are a major group of wood-destroying insects that we will discuss in the following chapter. In this section, we will discuss the three main groups of wood borers: ambrosia beetles (Scolytidae, Platypodidae), phloem borers (Cerambycidae, Buprestidae), and powderpost beetles (Bostrichidae, Lyctidae). We will also discuss a fourth group of unique freshwater wood borers that occur in Ghana (Polymitarcidae). Marine wood borers are important pests of wood in contact with salt water but, because they are not insects, they are beyond the scope of this book.

Ambrosia Beetles

Ambrosia beetles are a unique group of wood-boring insects that belong to the families Scolytidae and Platypodidae. These beetles bore into wood and inoculate symbiotic association with fungi. The beetles bore into wood and inoculate their tunnels with fungi; these fungi are often carried in a specialized structure on the beetle called a mycangium (pl.; mycangia). The beetle gallery is an ideal environment within which the fungus flourishes. The beetles get their nourishment by feeding on the fungus, not on the wood. When this phenomenon was first observed in the late 1800s, it was considered so unique that the term "ambrosia" (literally "food of the gods") was used to describe this relationship. The term ambrosia beetle has been used ever since to refer to this group of wood-boring insects. There are roughly 200 species of ambrosia beetles in Ghana (Browne 1963).

Because there are some aspects unique to each of the two families, Scolytidae and Platypodidae, we will discuss the life history of each family separately and provide one specific example for each family. Then we will discuss the damage and control for the two families combined.

Scolytidae: Taxonomy and Biology

The Scolytidae is a large family of Coleoptera often referred to as the bark and ambrosia beetles (Figure 6.1). There are numerous species of bark beetles that occur in Ghana. These beetles generally confine their boring to the bark and rarely enter the wood. Consequently, they are of little or no economic importance to Ghana forests. It should, however, be noted that bark beetles, especially those in the genera *Dendroctonus, Scolytus*, and *Ips*, are major forest pests of the coniferous forests of the world. These insects are capable of causing widespread mortality, and could become a serious problem if pines were widely planted in Ghana and these insects were introduced.

The ambrosia beetles within the Scolytidae are rather short, rounded beetles and are brown to black in color. The beetles are small, rarely exceeding 1 cm in length. These ambrosia beetles have quite variable habits. Some species, such as *Rhopalopselion thompsonii, Scolytoplatypus acuminatus*, and *Dactylipalpus camerunus*, are monogamous in habit while others such as *Stylotentus concolor, Premnobius sexspinosus*, and *Xyleborus affinis*, are extremely polygamous. Some species, like *Hypothenemus camerunus* and *S. concolor*, attack small branches and twigs while others, like *D. camerunus* and *X. affinis*, prefer moderate-sized host material. Gallery patterns can be quite variable. In many of the ambrosia beetles in this group, females are well developed and males are reduced, sometimes wingless, and occasionally completely absent. *Eccoptopterys sexspinosus*, for example, has only a few small, wingless males. In general, the female chooses the host and is followed by the male whose primary purpose is mating. Females construct the entire gallery system.

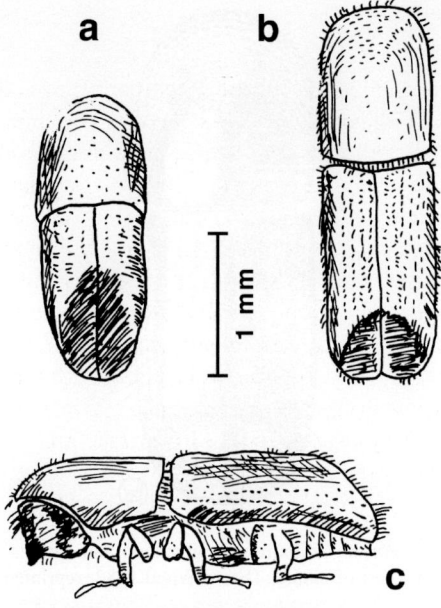

Figure 6.1 *Premnobius cavipennis*, a typical ambrosia beetle in the family Scolytidae: (a) male (dorsal view); (b) female (dorsal view); (c) female lateral view. (Redrawn from Browne 1961a, b.)

A Typical Scolytid Beetle: *Xyleborus ferrugineus*

Xyleborus ferrugineus is an extremely cosmopolitan species of Scolytidae that exhibits the ambrosia habit (Figure 6.2). This species occurs throughout the tropical areas of Africa, Madagascar, southern USA, Central America, South America, and the South Pacific. This species occurs in 74 host plants representing 29 families (Browne 1962a, b). *Xyleborus ferrgineus* is a small, abundant species that occurs throughout the high forest zone of Ghana. It normally breeds in fallen logs, dying trees, and fresh-cut logs with or without bark. This species has also been reported to attack living trees through wounds that presumably go to the xylem tissue. There are a few reports of this species occurring in woody seeds (Browne 1962a, b).

As is typical for this genus, *X. ferrugineus* is polygamous with flightless males that are reduced in size and number. Males do not normally leave the parent nest. Adult females fly year round (Atuahene and Nkrumah 1977) and are crepuscular or nocturnal in flight habit. Females choose the host tree and initiate gallery formation. Galleries tend to be short (less than 5 cm). Tunnels are irregularly branched, generally in one transverse plane of the wood (Browne 1962a, b). The life cycle of this insect in Ghana is not well known, but development from egg to adult requires 4–6 weeks. Because this species is abundant and breeds year round, it has a high-potential pest status in fresh-cut logs.

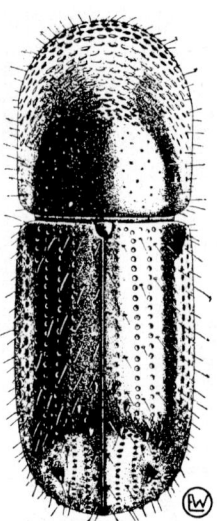

Figure 6.2 *Xyleborus ferrugineus*, an extremely cosmopolitan species of ambrosia beetle that occurs throughout the high forest of Ghana. (From Wood 1982; reprinted with permission.)

Platypodidae: Taxonomy and Biology

The Platypodidae is a smaller group than the Scolytidae, but still has many important species. In general, the Platypodidae are much less variable than the Scolytidae. All are monogamous and ambrosia beetles. Males and females are equal in size, and both function in host selection and gallery construction. Sexes are normally separated on the basis of head and elytral ornaments. These beetles are similar in size to Scolytidae, except the body is elongate and cylindrical in shape.

In general, the males select the host and initiate the gallery in advance of the female. The female joins the male later and is responsible for the majority of the gallery construction. Males function to remove debris from the gallery as the female bores and also guards the tunnel entrance. The gallery pattern is variable. For example, *Periommatus exisus, Cylindropalpus affinis*, and *Platypus hintzi* all bore galleries in one plane, while *Doliopygus conradti* bores several transverse planes. In general, platypodid bore much deeper galleries into wood and prefer larger sized host than do the scolytids. Two species are important borers of living trees, *Doliopygus dubius* and *Trachyostus ghanaensis*, and were discussed in the previous section on wood borers of living trees.

A Typical Platypodid Beetle; *Doliopygus conradti*

Doliopygus conradti is a very common platypodid that occurs throughout the forest areas of Ghana and is widely distributed in West Africa (Figure 6.3). The beetle is one of the larger Platypodidae and attacks a large number of unrelated hosts. It

Figure 6.3 White boring dust at the entrance of an ambrosia beetle gallery

normally breeds in relatively large dead or dying stems. In Ghana, *D. conradti* has been reported attacking apparently healthy trees including the economically valuable *Entandrophragma angolense*. These attacks are rarely successful, but damage and degrade remains in the log (Browne 1968).

Adults are crepuscular and both sexes fly at the same time. Browne (1968) reports continuous breeding year round; Atuahene and Nkrumah (1977) observed adult flight only between September and March. Males initiate attack of the tree and bore a short gallery. When females alight on a log they search for males, which will only mate after their tunneling response is satisfied. Females mate before they begin tunneling (Browne 1962a, b). Once mating is completed, the female constructs much of the nest with the male assisting by keeping the gallery free of frass. Several clutches of eggs are laid and the first adults leave the nest approximately 7 weeks; emergence is irregular and continues for a prolonged period (Browne 1968).

The tunnels penetrate deeply into the wood and staining is prevalent. Tunnels lie in several transverse planes that are connected by longitudinal shafts. Pupal cells are grouped alone and below transverse auxiliary galleries (Browne 1968).

Because this species is quite abundant and prefers larger sized material, it has the potential to be a serious economic pest. The ability to attack living trees amplifies the potential importance of this species.

Ambrosia Beetle Damage (Scolytidae, Platypodidae)

A distinct indication that ambrosia beetles are present in logs is small piles of white boring dust at the entrance to the tunnels (Figure 6.3). When beetle populations are

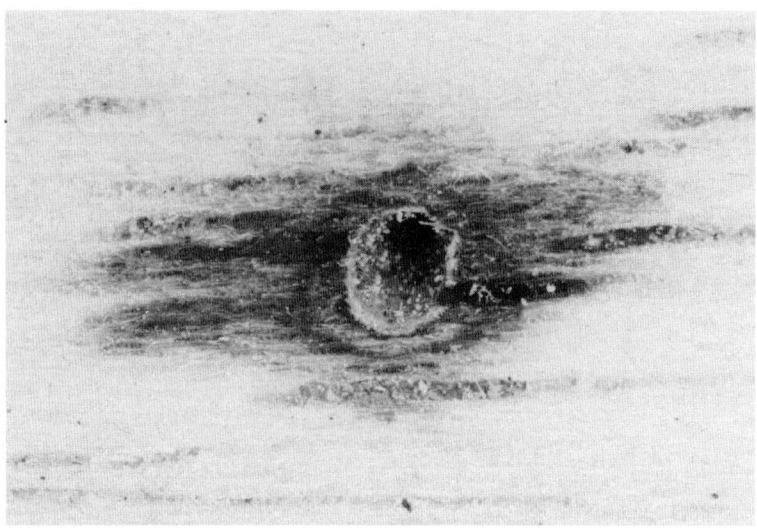

Figure 6.4 Enlarged ambrosia beetle gallery illustrating the degree of wood staining around the gallery and showing ambrosia beetle fungi growing on the walls of the tunnel

high these piles of dust can accumulate to a considerable degree around logs (Plate 53). In lumber, the damage results in a distinctive attack pattern. Infested lumber is riddled with small holes that generally have dark staining from the ambrosia fungus (Figures 6.4 and 6.5, Plate 54). Occasionally the saw cut will dissect a gallery longitudinally, leaving an oblong gallery in the end of the log. Both the cross-cut end and rounded surfaces of billets can be attacked, but the attacks to the end grains are rarely successful. Most ambrosia beetles attack logs within days or weeks of felling. However, some species prefer logs that have been in the bush for several weeks. Many of the ambrosia beetles prefer logs without bark intact (Anonymous 1957). Wood damage by ambrosia beetle attack is less marketable because it is riddled with galleries and stained. The presence of pin holes results in rejection of logs from Ghana. This is an unnecessary practice, because ambrosia beetles rarely infest the heartwood (Figure 6.5). In most tropical hardwoods the sapwood is culled during conversion anyway. The practice of removing sapwood will likely be halted in the future because trees of smaller diameter and a greater proportion of sapwood will be increasingly utilized. However, buyers are still reluctant to buy logs or lumber infested with ambrosia beetles. Structural strength of ambrosia beetle-infested wood is equal to sound wood. Ambrosia beetle-damaged wood that has been dried can be used quite effectively for uses where the wood remains dry. The presence of holes and stain modifies the permeability characteristics of the wood which results in differential stain or paint absorption. In structural uses where appearances are not important, ambrosia beetle damage is not a significant degradation.

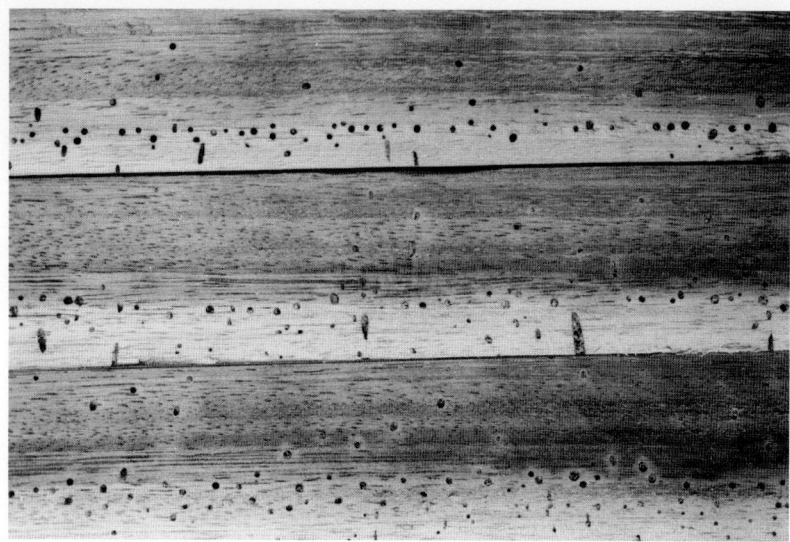

Figure 6.5 Ambrosia beetle damage to three strips of odum wood. Note the heavier attack rate in the lighter-colored sapwood

Usually, most tree species are susceptible to ambrosia beetle attack. There can, however, be considerable variation in the degree and rate of attack. Some species, like *Albizia africana* and *Antiaris africana*, are attacked quickly and in very high density. Others, like *Turraenthus vignei* and *Cistanthera papaverifera*, are relatively less susceptible to attack (Anonymous 1957). The relative susceptibility of 20 tree species is provided in Table 6.1.

Pest Management

The best method of control for ambrosia beetle is quick removal of logs from the forest and rapid conversion. It usually takes a few days for the beetle to get established and quick removal will minimize damage. However, quick removal of logs and rapid conversion are often difficult and expensive. The presence of tree bark will retard ambrosia beetle damage for a short time; however, the presence of bark encourages attack by longhorn beetles. If logs are to remain in the forest for a week or more, the logs should be peeled and sprayed with an insecticide (Plate 55). In other parts of the world these insects can be controlled by felling during periods when the insects are not active. Because of the year-round activity of most Ghanaian species this technique is not effective in the tropics. The ambrosia beetles that occur in Ghana are summarized in Table 6.2.

Table 6.1 Relative susceptibility of 20 species of Ghana trees to attack by ambrosia beetles within 6 weeks of felling. (Modified from Anonymous 1957)

Species	Susceptibility rating
Turraenthus vignei	Low
Cistanthera papaverifera	Low
Piptadeniastrum africana	Moderate
Entandrophragma utile	Moderate
Entandrophragma cylindricum	Moderate
Mansonia altissima	Moderate
Terminalia ivorensis	Moderate
Guarea cedrata	Moderate
Entandrophragma angolense	High
Entandrophragma candollei	High
Terminalia superba	High
Khaya ivorensis	High
Triplochiton scleroxylon	High
Celtis zenkeri	High
Cylicodiscus gabunensis	High
Nauclea diderrichii	Very high
Pycnanthus angolensis	Very high
Milicia excelsa	Very high
Albizia ferruginea	Extreme
Antiaris africana	Extreme

Phloem Borers

Phloem borers received their name from their habit of feeding on the phloem tissue of trees. Most members of this group lay their eggs beneath or within the bark of their hosts. The young larvae feed primarily in the phloem tissue. As the larvae mature and grow in size, they begin to feed in the interface between the phloem and the xylem. At this time etching of the xylem occurs and many species will bore completely into the xylem to complete development and pupate. Some species will spend considerable time feeding in the xylem tissue, which is where the greatest economic damage results. The insects in this group are Coleoptera in the families Cerambycidae and Buprestidae. Both groups are quite large with considerable variation in life history strategies. We will discuss each family separately and restrict our discussion to those species that have the wood boring habit.

Longhorn Beetles (Cerambycidae)

Longhorn beetles, like ambrosia beetles, are worldwide in distribution and abundant in West Africa. There are 15,000 species varying in size from less than 2 cm

Table 6.2 Ambrosia beetles in Ghanaian forests

Species	Host plant	Family preference	Life history/comments
Coleoptera: Platypodidae			
Chaetastus tuberculatus Chapl;	*Ricinodendron heudelotii, Milicia excelsa, Entandrophragma angolense, Entandrophragma candollei, Albizia zygia, Ficus exaeperata, Khaya grandifoliola, Terminalia ivorensis, Entandrophragma cylindricum, Khaya ivorensis, Mitragyna atipulosa, Triplochiton scleroxylon*		Occurs throughout Africa south of Sahara; a large beetle breeding in logs or stressed trees; gregarious species resulting in scattered attacks; tunnels on transverse plain of wood; adults fly in January
Cylindropalpus affinis	*Sapotaceae*		
Cylindropalpus auricomans Schauf	Not known in Ghana		Large trees and branches attacked; common species; galleries form semicircular tunnel in transverse plane; short vertical tunnels and in pupal chambers; throughout high forests of Ghana; adults fly in September–November
Cylindropalpus camerunus			Biology and tunneling similar to C. auricomans; adults fly in October
Cylindropalpus pumilio	*Ficus exasperata*		Small species; not common; adults fly in June, October–November; adults attacked to light
Diapus guinguespinatus Chap	No host preference		Small species; galleries in one plane; galleries penetrate heartwood; important problem in the Orient
Doliopygus angolensis			Found throughout high forest zone, adults fly in September–March
Doliopygus aduncus	*Daniellia oliveri*		In Nigeria adults are observed often boring into living trees; dead trees favored for breeding; savannah species
Doliopygus chapuisi Duv	*Albizia adianthifolia, ferruginea, Albizia glaberrima, A. zygia, T. scleroxylon, Cedrela mexicana, Cassia siamea*	*Mimosaceae*	Large ambrosia beetle; crepuscular flight; deep penetrating gallery; monogamous; branched galleries in the transverse plane of the wood; species active at all times of the year

(continued)

Table 6.2 (continued)

Species	Host plant	Family preference	Life history/comments
Doliopygus coelocephalus Schauf	*A. zygia, Celtis mildbraedii*		Large species; common at lights in early evening
Doliopygus conradti Strohm	*A. glaberrima, A. zygia, Antiaris africana, Blighia sapida, Bussea occidentalis, Carapa procera, Celtis brownie, C. mildbraedii, M. excelsa, Chrysophyllum albidium, Cleistopholis patens, Cola acuminata, Cylicodiscus gabunensis, E. angolense, E. cylindricum, Entandrophragma utile, Guarea cendrata, Hannoa klaineana, Hymenostegia afzelii, K. ivorensis, Mimusops heckelii, Morus mesozygia, Musanga cecropioides, Nauclea diderrichii, Mansonia altissima, Panda oleosa, Piptadeniastrum africanum, Pterygota macrocarpa, Pycanthus angolensis, R. heudelotii, Sterculia tragacantha, Strombosia glaucescens, Terminalia superba, T. scleroxylon, Trichilia heudelotii*	Melicaea, Moraceae	Very common through high forest zone; crepuscular adults fly in September–March.; branch galleries each with larval cradles from entrance hole horizontal, vertical or spiral; infest logs <1 ft; galleries deep in wood
Doliopygus dubius Samps	*A. glaberrima, C. mildbraedii, Chrysophyllum albidum, E. angolense, E. cylindricum, K. ivorensis, M. mesozygia, P. angolensis, T. superba, T. scleroxylon, Anthocleista nobilis, B. sapida, C. mexicana, Ficus exasperate, P. africanum, T. ivorensis*		A large ambrosia beetle common breeding in logs and dying trees; monogamous; nest construction initiated by male, then female, and completed by larvae; deep penetration of wood; all stages present throughout the year occurs throughout the high forest zone
Doliopygus erichsoni Chap	*Spathodea campanulata*, wide host range including *Tectona grandis* in Nigeria		Large ambrosia beetle; crepuscular flight; continuous breeding; can attack healthy trees at the onset of dry season; galleries penetrate wood deeply and are branched

Species	Hosts	Family	Biology
Doliopygus exilis Chap	*A. zygia, Bosqueia angolensis, H. afzelii, C. mildbraedii, P. africanum, R. heudelotii*	*Mimosaceae*	Attacks felled or windblown trees; entrance tunnels occur in small groups; found throughout high forest zone, adults fly in October–March
Doliopygus falcivicus	Wide host range		Attacks moderate to large felled trees; adults active mid–late morning; adults fly in April–May, December
Doliopygus gracilior Schedl	*A. zygia, P. africanum*		Small species; attacks newly felled trees; attacks trees quickly; crepuscular adult flight
Doliopygus interjectus Schedl	*Baphia nitida, T. heudelotii, T. scleroxylon*		Attacks felled trees of small to moderate size; tunnels enter heartwood of Trhe; adults tunnel in February, September
Doliopygus interpositus Schedl	Wide host range		Widespread species; attacks wide size range of hosts; diurnal adult flight
Doliopygus malkini Schedl	*Mimosaceae*	*Mimosaceae*	Savannah species; bark covered areas of newly felled, moderate to large trees preferred; gallery in one plane; adult–adult 5–8 weeks
Doliopygus minutissimus			
Doliopygus nairobiensis Schedl	*A. africana, E. cylindricum, B. nitida*		Crepuscular flight; adults tunnel in February
Doliopygus notatus Schedl	*T. scleroxylon*		
Doliopygus opifex Samps	*A. zygia, A. africana, B. occidentalis, H. afzelii, P. oleosa, P. africanum, Afzelia africana, T. scleroxylon*		Uncommon species
Doliopygus perbrevis Schedl	*A. zygia, A. Africana, B. sapida, C. mildbraedii, C. gabunensis, E. utile, G. cendrata, H. afzelii, M. mesozygia, P. oleosa, S. glaucescens, T. superba, T. heudelotii, B. angolensis*		Common species that attacks recently felled trees of all sizes; adults tunnel in January–March, November
Doliopygus perminutissimus Schedl	*A. zygia, P. africanum, C. siamea*		Entrance tunnel in dead trees confined to dried bark; tunnel in *C. siamea*, enters heartwood
Doliopygus piptadeniae Schedl	*A. zygia*		

(continued)

Table 6.2 (continued)

Species	Host plant	Family preference	Life history/comments
Doliopygus propinguus Schedl	*Albizia ferruginea, A. zygia, Cola nitida, C. mildbraedii, E. utile, H. afzelii, Mammea africana, Piptadenastrum africanum, S. glaucescens*	Mimosaceae	Adults fly in December, attacks restricted to areas of dry bark
Doliopygus punctiventris Schedl	*A. africana*		
Doliopygus rapax Schedl	*B. angolensis, F. exasperate, P. macrocarpa*		Small common species found throughout high forest zone; adults fly in September–July
Doliopygus retusus			
Doliopygus serratus Strohm	*A. zygia, C. nitida, Turraeanthus vignei, B. sapida, Celtis adolfi-friderici, C. mildbraedii, Ficus, P. macrocarpa, T. ivorensis, T. heudelotii, T. scleroxylon*		Widely distributed throughout the forest zone; normally breeds in logs and dying trees but can attack healthy trees including Sterculia rhinopetala in Ghana; monogamous; galleries penetrate deeply in wood; continuous breeding; prefer logs with bark intact; attacks all sizes of logs; attacks cut logs quickly
Doliopygus spinosus			Crepuscular flight
Doliopygus tenius Strohm	*C. brownie, C. mildbraedii*	Ulmaceae	Crepuscular flight
Doliopygus umbonatus			Adults fly in April–May, December–January
Doliopybus unispinosus Schedl	*A. zygia, B. occidentalis, Celtis brownii, C. mildbraedii, P. macrocarpa, E. utile, T. scleroxylon, Azadirachta indica*		Normally breeds in fallen trees; found throughout high forest zone; adults fly in April–May, November–December monogamous deep branched galleries; adults attack logs quickly
Periommatus angustiformis Schedl			Periommatus genus of slender ambrosia beetles; abundant year round species throughout high forest zone; monogamous; branched galleries in transverse plane of wood; prefers trees that have been felled for a while

Species	Host plants	Plant families	Biology
Periommatus angustior Schedl	*Funtumia elastica*		Abundant throughout high forest zone
Periommatus camerunus Strohm	*C. mildbraedii*		Throughout high forest zone; adults fly in January–June, September–October
Periommatus excisus Strohm.	*A. africana, C. mildbraedii, Guarea cedrata, E. angolense, T. ivorensis, B. angolensis, F. exasperate, T. scleroxylon*		Throughout high forest zone; adults fly in January–May; adults nocturnal
Periommatus grandis Schedl	*C. brownie, C. mildbraedii, C. albidum, Ficus vogelii, Maesobotrya bateri, P. africanum, Trichilia heudelotti, Albizia globerrima, B. sapida, T. scleroxylon*		Adults nocturnal
Periommatus mkussi			Throughout high forest zone; adults fly in January, June–September, November
Periommatus pseudomajor Schedl	*F. exasperata*		Found throughout high forest zone; adults fly in January–March, May
Periommatus substriatus Strohm			
Platypus augustatus Strohm	*A. africana, C. brownie, C. mildbraedii, M. excelsa, M. bateri, Pycnanthus angolensis*		
Platypus erichsoni Chap	*T. scleroxylon*		
Platypus hintzi Schauf.	*A. zygia, A. glaberrima, A. africana, B. angolensis, C. brownie, C. mildbraedii, Celtis zenkeri, Chlorophora excelsa, C. albidum, C. patens, Cola gigantean, C. gabunensis, E. angolense, E. cylindricum, E. utile, Ficus asperifolia, F. vogelii, H. klaineana, Musanga cecropoides, N. diderrichii, P. angolensis, Trichilia neudelotti, T. scleroxylon, Alstonia boonei, A. nobilis, C. mexicana, F. exasperate, Lonchocarpus sericeus, P. macrocarpa, R. heudelotii*	Mimosaceae, Meliaceae, Moraceae	Slender beetle; abundant year round throughout high forest zone; adults crepuscular; infests sawn boards; monogamous; may be the most abundant and most polyphagous ambrosia beetle in Ghana; tunnels in radial plane; pupal chambers in groups of 4–5.

(continued)

Table 6.2 (continued)

Species	Host plant	Family preference	Life history/comments
Platypus impressus	Wide range		Small platypodid; adults fly in October, December; occur in partially dried logs; galleries similar to *P. hintzi*; adults diurnal
Platypus intermedius Strohm	*A. africana, Cercestis afzelii*		Prefer logs with bark intact
Platypus mordax Samps	*A. zygia, T. scleroxylon, Conopharyngea* sp.		
Platypus orientalis Strohm	*A. glaberrima, A. africana, F. exasperate, Futumia elastica, P. africanum*		Attacks windblown or felled trees; on partially green dead trees; entrance holes are confined to areas of dry bark; deep extensive tunnel; adults tunnel in October–December
Platypus pygmaeus Schedl	*A. zygia, A. africana, H. afzelii, F. exasperata*		
Platypus spinulosus Strohm Platypus refescens	*A. africana* (Moraceae)		Attack logs and drying tree with or without bark intact Adults fly in November
Platyscapus auricomus Schedl	*B. occidentalis, C. nitida, Combretodendron africanum, C. albidum, Distemonanthus benthamianus, E. utile, H. klaineana, P. macrocarpa, T. heudelotti, T. scleroxylon, Chrysophyllum* sp.		Galleries have been observed in small branches penetrating to the pith; gallery with circumferential and longitudinal branches
Platyscapus camerunus Schedl	*A. africana, T. scleroxylon*		
Platyscapus pumilio Schedl	*T. scleroxylon*		
Trachyostus aterrimus (Schauf.)	*C. adolfi-friderici, C. mildbraedii, C. zenkeri, C. gigantean, Sterculia rhinopetala, H. afzelii, Afzelia africana, P. macrocarpa, A. ferruginea, T. superba*		Nest formation similar to *Trachyostus ghanaensis;* monogamous; only attacks logs several months old; adults diurnal; attacks stems of living trees; extensive damage throughout Ghana
T. ghanaensis Schedl	*A. africana, T. scleroxylon*		Details in text
Trachyostus carinatus Schedl	*Ricinodendron heuelotii*		New species from Ashanti
Trachyostus schaufussi Schedl	*E. angolense, E. cylindricum, A. glaberrima, A. zygia, C. mildbraedii, A. africana, B. nitida*		Prefers logs with bark, adults diurnal; deep penetrating gallery; prefers large trees
Triozastus elongates Schedl	*A. zygia, P. africanum*		Small uncommon species in felled trees

Species	Host plants		Notes
Triozastus marshalli			Rare in Nigeria; prefers felled logs; common year-round species throughout high forest zone in Ghana
Triozastus philosulus			Gallery system consists of semicircular tunnel in tunnel in transverse plane; adults fly in November–December, entrance gallery with projecting cylinder of latex
Triozastus propatulus Schedl	*A. ferruginea, A. zygia, A. africana, B. angolensis, B. nitida, P. africanum, T. heudelotti, T. scleroxylon*		
Coleoptera: Scolytidae			
Coccotrypes congonus			
Coccotrypes rutshuruensis Egg	*T. scleroxylon*		
Cryphalomorphus sp.			Adults fly in June
Cryphalomorphus ghanaensis (Schedl)	*H. afzelii, B. occidentalis*		Adults fly in November New species from Ashanti monogamous
Cryptocarenus heveae (Hag.)	*Mangifera indica, C. siamea*		Extreme polygamy: males are few; attack small twigs; irregular longitudinal galleries in pith
Crytogenius cribicollis	*Elaeophorbia drupifera*		Found only on Accra Plain; monogamous; attacks stems and branches of dead and dying trees; occasionally attacks living trees in dry season
Dactylipalpus camerunus (Hag.)	*T. superba, C. mildbraedii*		Uncommon species in the southern high forest; prefers moderate sized material
Eccoptopterus sexspinosus Motsch.	*A. zygia, B. angolensis, A. adianthifolia*	Mimosaceae	Mimosaceae branchwood preferred; only occurs on logs devoid of bark; adults diurnal; radial entrance gallery; breeding into short secondary tunnels in various planes of the wood
Hypothenemus camerunus (Egg.)	*Ceiba pentandra, P. africanum, B. nitida, Baphia pubescens, S. tragacantha, T. scleroxylon, Trema guineensis*		Occurs mainly on small branches and twigs; often only in bark but also will bore all the way to the pith; irregular galleries

(continued)

Table 6.2 (continued)

Species	Host plant	Family preference	Life history/comments
Hypothenemus socialis	*B. occidentalis, C. siamea, Ficus asperifolia, T. scleroxylon, C. mildbraedii, Trema guineensis*		Similar to *H. camerunus*; also bore to pith
Mimips giconicus Schedl	*P. macrocarpa*		Found in branchwood of *Pterygote macrocarpa* in Nigeria; attracted to lights at night
Premnobius ambitiosus Schauf	*A. zygia, C. mildbraedii, Gmelina* sp.		Adults fly in August–November, January–June; throughout high forest zone; short entrance tunnel with short side tunnels in the tangential plane
Premnobius cavipennis Eichh	*C. patens, Distemonanothus benthamianus, H. afzelii, T. ivorensis, T. superba, Guarea cedrata, T. heudelotti, A. adianthifolia, A. zygia, C. gabunensis, A. africana, M. mesozygia, P. oleosa, B. sapida, C. albidum, P. macrocarpa, Sterculia rhinopetala, T. scleroxylon, C. mildbraedii, C. siamea*		Prefers branchwood (fresh or dead); very common throughout high forest zone, also occurs in tropical America; extremely polygamous; host selection by female; simple branched tunnel <3 cm into log; eggs laid in clusters
Premnobius corthyloides Hag.	*C. siamea*		Adults fly in October–May; throughout high forest zone
Premnobius longus	*C. siamea*		Less common than other species of the genus; adult fly in September–October, January–May; throughout high forest zone
Premnobius minor			Adults fly in September
Premnobius sexspinosus Egg			Adult females attracted to light; adults fly in September, October
Premnobius xylocranellus Schedl	*A. zygia, A. africana, B. sapida, C. mildbraedii, C. albidum, Cleistopholia patens, C. gabunensis, Distemonanthus benthamianus, Guarea cedrata, H. afzelii, M. mesozygia, P. oleosa, P. macrocarpa, T. scleroxylon*		Gallery pattern and biology similar to *P. ambitiosus*
Rhopalopselion confusum (Egg)	*B. angolensis, Baphia* spp.		

Species	Host	Biology
Rhopalopselion thompsoni Schedl	*B. occidentalis, D. benthamianus, B. nitida, B. pubescens, Cola caricifolia, C. mildbraedii*	Species known only to Ghana; monogamous; occasionally in branches; tunnels short, 1–2cm
Scolytoplatypus acuminatus Schedl	*B. occidentalis, Trichilia lanata*	Occurs on branchwood in high forest, monogamous; simple branched galleries
Scolytoplatypus occidentalis	Members of this genus have broad host ranges	Scolytoplatypus galleries generally have regularly branched curved tunnel in the transverse of the wood; breeds continuously
Streptocranus adjanctus		Adults fly in January, June
Streptocranus usugaricus		Adults fly in January
Strombophorus ericius (Schauf.)	*C. brownii, C. mildbraedii, C. zenkeri*	Attacks branchwood 1–12cm in diameter; short galleries into food; feeds in bark and wood
Stylotentus concolor		
Tiaphorus camerunus		
Tiaphorus elongates		Adults fly in June, December; throughout high forest zone
Tiaphorus hypaspistes		Adults fly in January, September
Xyclocleptes brownie		
Xyleborus acanthus		
Xyleborus affinis Eichh.	*T. scleroxylon*	Very abundant species throughout high forest; year-round adult flight; normally breeds in dying or fallen trees; has attacked *Eucalyptus* in Nigeria
Xyleborus africanus		Adults fly in August–December; throughout high forest zone
Xyleborus albizzianus Schedl	*C. mildbraedii*	March adult flight
Xyleborus alluandi Schauf	Wide host range	Normally attacks felled trees or large branchwood in January–May, September–October diurnal adult flight; has been found living in *Eucalyptus*; throughout high forest zone; branched tunnels penetrate 2–5in

(continued)

Table 6.2 (continued)

Species	Host plant	Family preference	Life history/comments
Xyleborus ambasius Hag.	*A. ferruginea, A. africana, C. gabunensis, E. cylindricum, K. ivorensis, N. diderrichii, P. angolensis*		Adults fly in May–June
Xyleborus ambasiusculus Egg	*A. ferruginea, A. africana, G. cedrata, K. ivorensis, M. altissima, N. diderrichii, T. superba, T. scleroxylon, B. sapida, T. ivorensis*		Small beetle
Xyleborus badius Eichh.	*A. ferruginea, A. africana, A. zygia, A. nobilis, C. siamea, Milicia excelsa, F. exasperate, P. africanum, Ricinodendron heudelotti, T. ivorensis, T. heudelotti, T. scleroxylon*		Adults attack recently dead or fire killed trees; entrance at right angle to bark, deep into sapwood usually not in heartwood; many branched gallery, no consistent pattern; uncommon species; adults fly in May–June, December
Xyleborus barumbuensis Egg. *Xyleborus cameranus* Hag.	*A. ferruginea, A. glaberrima, A. zygia, A. africana, C. brownie, C. zenkeri, C. mildbraedii, Milicia excelsa, C. gabunensis, Distemonanathus benthamianus, E. angolense, E. cylindricum, E. candollei, E. utile, H. afzelii, K. ivorensis, M. altissima, M. mesozygia, N. diderrichii, P. africanum, P. angolensis, R. heudelotti, S. glaucescens, Tarrietia utilis, T. superba, T. scleroxylon, T. vignei*	Meliaceae, Mimosaceae	Galleries with simple branches radical in one plane from entrance; no larval cradles; adults crepuscular
Xyleborus camphorae Hag. *Xyleborus confuses* Eichh.	*Celtis zygia* *A. zygia, C. brownie, C. mildbraedii, M. excelsa, C. gabunensis, E. angolense, T. superba, T. scleroxylon, A. nobilis, A. africana, Cussonia barteri, P. macrocarpa, R. heudelotti*	Ulmaceae	Prefers logs with bark intact Biology similar to *X. badius*
Xyleborus collarti Egg	*T. heudelotti, Conopharyngea* sp.		Prefers felled trees and branchwood

Species	Host plants	Family	Biology
Xyleborus conradti Hag.			Adults fly in January–February, June–December; throughout high forest zone
Xyleborus cristatus (F.)			Fairly common; diurnal flight
Xyleborus diversus Schedl	*Athocleista nobilis, T. scleroxylon*		
Xyleborus eichhoffi Schr.			Newly felled and dying trees attacked; adults crepuscular; Adults fly in October–December; throughout high forest zone
Xyleborus eichoffianus Schedl	*C. mildbraedii, Mitragyna stipulosa, A. zygia, F. exasperate, T. ivorensis, T. scleroxylon, A. ferruginea, A. africana, C. zenkeri, Milicia excelsa, E. candollei, M. altissima, T. scleroxylon*		
Xyleborus ferrugineus F.:	*Albizia ferruginia, A. africana, C. zenkeri, Milicia excelsa, E. candollei, M. altissima, T. scleroxylon*		Galleries irregularly branched 3–8 cm inward; abundant species throughout high forest zone; year round adult flight; logs and large branches commonly attacked; adults nocturnal; confined to sapwood; distributed throughout tropics of the world; males incapable of flight and few in number
Xyleborus indicus Eichh.	*A. glaberrima, A. africana, C. brownie, C. mildbraedii, Celtis zygia, M. excelsa, C. gabunensis, E. angolense, E. cylindricum, K. ivorensis, M. altissima, M. stipulosa, N. diderrichii, P. angolensis, T. ivorensis, T. scleroxylon, T. vignei*	Meliaceae Ulmaceae	Adults fly in May–June, December; throughout high forest zone; adults diurnal; confined to sapwood
Xyleborus mascarensis Eichh.	*A. ferruginea, A. zygia, A. africana, Anopyxis klaineana, C. procera, C. brownie, C. mildbraedii, C. zenkeri, M. excelsa, C. gabunensis, Distemonanathus benthamianus, E. angolense, E. candollei, E. cylindricum, E. utile, Guerrea cedrata, K. ivorensis, M. altissima, M. heckelii, M. stipulosa,*	Meliaceae, Sterculiaceae	Adults crepuscular; confined to sapwood; observed infesting cut boards; can attack living saplings

(continued)

Table 6.2 (continued)

Species	Host plant	Family preference	Life history/comments
	M. cecropioides, N. diderrichii, Nesogordonia papaverifeva, M. africana, P. africanum, P. angolensis, Sterculia rhinopetala, S. glaucescens, Tarrieta utilis, T. ivorensis, T. superba, T. heudelotti, T. scleroxylon, Turraeanthus vignei		
Xyleborus neogranulatus	T. scleroxylon		
Xyleborus perforans Woll.	Unselective in host, > 100 sp. As hosts		Usually breeds in newly cut logs; adults fly in February–May, June–November; readily attacks newly sawn lumber; throughout high forest zone; polygamous species with few males which are incapable of flight; irregular tunnel without communal chambers
Xyleborus picinus			Adults fly in August–September
Xyleborus perdililigens Subsp. diligens Schedl	A. zygia		
Xyleborus psaltes	A. zygia		Adults fly in August–September; throughout high forest zone
Xyleborus pseudoembasius	T. scleroxylon		Adults fly in August–September
Xyleborus ricini Egg.	Albizia spp.		September adult flight throughout high forest zone; attracted to lights at night
Xyleborus scabrior Schedl	T. scleroxylon		
Xyleborus scabrior Schedl	T. scleroxylon		
Xyleborus semiopacus Eichh	A. ferruginea, A. zygia, A. africana, C. procera, C. brownie, C. mildbraedii, C. zenkeri, C. excelsa, C. patens, C. gabunensis, D. benthamianus, E. angolense, E. cylindricum, E. utile, Guerra cedrata, K. ivorensis, M. stipu		Small beetle that breed in small stems usually confined to sapwood; can cause serious damage to new plantations of Aucoumea klaineana, K. ivorensis; forms one or more large communal galleries near entrance where all stages are found; adults crepuscular; adults

	losa, Nauclea diderrichi, P. africanum, T. superba, T. scleroxylon, T. heudelotti	fly in January, August–September in Ghana, continuous in other areas; occurs throughout the high forest areas
Xyleborus sharpae Hop.	*A. ferruginea, E. utile, K. ivorensis, A. zygia, P. africanum, T. heudelotti*	Adults diurnal; can attack living saplings; biology similar to *X. badius*
Xyleborus similans Egg.	*Funtamia elastica, T. scleroxylon*	
Xyleborus subtuberculatus		Adults fly in year-round; abundant species throughout high forest zone
Xyleborus solitarius Hag.	*T. scleroxylon*	
Xyleborus torguatus Egg.	Wide host range in Nigeria	Galleries simply branched, normally penetrating only the sapwood in one plane; fallen trees, dying trees, and branchwood attacked
Xyleborus tropicus Hag	*B. occidentalis, T. superba, Sterculia rhinopetala, T. scleroxylon, C. mildbraedii*	Apparently confined to lowland rain forest
Xylosandrus compactus	*B. occidentalis, A. zygia, Ficus asperifolia, B. sapida, Trema guineensis*	Small ambrosia beetle that breeds in small material; polygamous with few males; parthenogenetic reproduction possible

Figure 6.6 *Ceroplesis quinquefasciata* is a typical Cerambycidae. The family gets its common name, longhorned wood borers, from the long antennae (adult body length 2.8 cm.)

to more than 10 cm (Jones 1959a, b). The common name describes the very long antennae which extend backward from the head and are often longer than the insect (Figure 6.6). Sometimes these insects are called round-headed wood borers because their galleries in cross section are generally round or oblong (no more than two times as long as wide).

Description and Life History

Adult longhorn beetles are elongate and often exhibit intricate patterns on the elytra. Like all Coleoptera, they have complete metamorphosis. The life cycle varies considerably in duration. In most of West Africa the life cycle is very short, approximately 4–6 weeks. There have, however, been examples of Cerambycidae and develops and emerges as long as 50 years following attack. The female lays eggs in

Figure 6.7 Typical evidence of feeding by longhorned wood borers. Note the tightly packed coarse frass in the galleries

or under the bark of trees using her sharp ovipositor. The first instar larvae are whitish in color, elongated, and clearly segmented. The head bears conspicuous dark, powerful mandibles. The larvae are legless. The young larvae bore actively beneath the bark, feeding on the wood and packing their tunnel with excrement and coarse wood fibers (Figure 6.7). The size of the tunnel corresponds to the diameter of the larvae. In Ghana, one species that attacks heartwood exceeds a diameter of 2.5 cm. Later larvae etch the wood surface and bore into the wood. The mature larva excavates a chamber in the wood for pupation (Figure 6.8). This pupal chamber is often located near the surface of the wood in a curved tunnel. The larva seals itself inside the chamber and pupates. The newly emerged adult bores out of the pupal chamber to the log surface.

Damage

Longhorn beetles normally confine their attack to unseasoned wood. Occasionally, however, they can be a problem in the dried lumber. The larvae are responsible for the damage, particularly when they bore into the wood to form pupation chambers (Figure 6.9). One exception is the adult *Analeptes trifasciata* which causes damage by ring-girdling stems of *Ceiba pentandra* in northern Ghana (Jones 1959a, b) and *Eucalyptus* spp. in southern Ghana (Atuahene 1975) (see previous section). Extensive damage is caused by some of the larger cerambycids that attack heartwood.

Figure 6.8 Mature longhorned wood borer larvae will often excavate tunnels into the wood

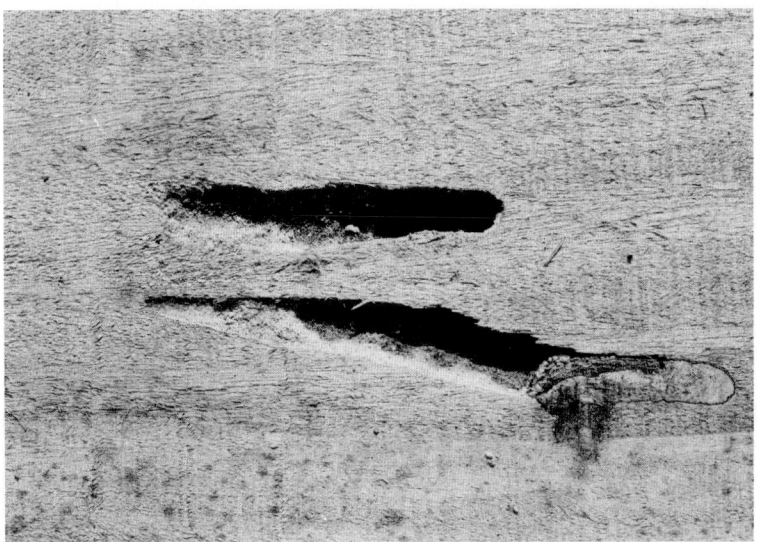

Figure 6.9 Cerambycid damage to lumber. This damage probably occurred prior to processing of the wood

Damage by longhorn beetles can generally be identified by the tightly packed coarse frass in the tunnels. The round to elliptical emergence hole is also helpful in identification.

Pest Management

Since the female lays her eggs in or beneath the bark, removing the bark will prevent attack. If the insect has already infested the log, quick conversion and culling of the sapwood would minimize damage. If the larvae are established in the heartwood, only processing and seasoning will stop further damage. The simplest method of control is to remove the bark. If logs must remain in the bush for some time before processing, preventative application of insecticides is appropriate.

Metallic or Flatheaded Wood Borers (Buprestidae)

This family in the order Coleoptera is mostly comprised of wood borers. Occasionally adults will feed on foliage and actually cause defoliation on forest trees. This family is commonly called the metallic or flatheaded wood borers.

Description and Life History

The general life history strategy of these insects is similar to that for the longhorn wood borers. Buprestids oviposit under the bark of dead or recently killed trees in oviposition niches cut by females. Some Buprestidae can be quite aggressive and attack living trees. *Agrilis hastulatus* is a typical bupresitid that attacks logs within as short a period as 48 h after felling. Though some species like *Agrilis portonovensis* prefer to oviposit in logs that have dried for some time.

All Buprestidae tend to mine the phloem tissue in the early stages of larval development. Later larval stages will begin etching the surface of the wood. Late larvae may bore deep into the sapwood where pupation normally occurs. The galleries tend to be packed with rather coarse boring frass. Many species like *A. hastulatus* complete their life cycle in approximately 8 weeks. In general, the phloem feeders have a longer developmental time than the ambrosia beetles because the ambrosia fungus has a higher food value than phloem tissue. *Chysobothris dorsata* has development time of 4 months, which is longer than many of the other Bupresitidae. Often the adults are metallic in color and the elytra end abruptly at the posterior end (Figure 6.10). The larvae are generally dorso-ventrally flattened with a distinctly widened second or third segment. Because the larvae are flattened, the holes they bore in wood are quite oblong (2–4 times as long as wide).

Figure 6.10 *Megactenodes westermanni* is a typical buprestid in Ghana

Damage

The damage caused by flatheaded wood borers is similar to that of longhorned wood bores. Larvae mine in the phloem and later etch the surface of the log. The larvae must bore into the wood to create pupation chambers. Most damage occurs in the sapwood. Bark must be intact for the log to be a suitable oviposition site. The damage to wood is typified by the very oblong galleries in cross section. The galleries are usually packed with frass and generally restricted to the sapwood.

Pest management

The recommendations for controlling these insects are identical to those for the Cerambycidae. The important coleopterous phloem borers of Ghana are summarized in Table 6.3.

Powderpost Beetle

The term "powderpost" adequately describes the type of damage which these insects cause. Heavily infested wood is a mass of powdery frass and the wood is structurally very week (Figure 6.11, Plate 56). Powderpost beetles are members of two distinct insect families; Bostrichidae and Lyctidae.

Bostrichidae

Bostrichids are common in the tropics of West Africa. There are 550 species known throughout the world. Though these insects normally attack wood that is old and has been processed into lumber, they can also occur in standing trees. For example, *Bostrychoplites productus* attacks standing trees near the base. These insects can create serious problems in completed wooden structures, particularly those that are old. Even well-seasoned wood that has been dried and kept dry can be attacked. The moderate tropical climate of Ghana makes these insects particularly important.

Description and life history

The beetles are generally small, ranging from less than 0.8 cm in length to more than 2.5 cm. One species in Ghana, *Apate terebrans*, is over 2.5 cm long and has a body diameter of approximately 1.2 cm. The adults are dark in color and elongate and cylindrical in shape. The head is always directed downward and covered from above by the hood-like thorax. This hood and the elytra are frequently sculptured and may bear small spines (Figure 6.12). Antennae are clubbed shaped and three segmented. Eyes and powerful jaws are present on the head.

The larvae are wrinkled, curved, and whitish in color. The head is inconspicuous, but possesses powerful jaws. The larvae have three pairs of legs and are active borers like the adult (Jones 1959a, b). The developmental period is quite variable among bostrichid species. The availability of starch is a key nutritional factor (Jones 1959a, b). Unlike ambrosia beetles, powderpost beetles actually feed on the wood. When starch content is low, insect development is relatively slow when compared to insects feeding on wood with high starch content. When environmental conditions are favorable the life cycle may be completed in 2–3 months. The female bores radially into the host for a short distance, then parallel to the host's surface to make a simple branched tunnel. Eggs are laid in the first branch (egg tunnel) where the larvae hatch and feed on the wood. After several moults, the larvae pupate. The emerging adult borers surface to complete the life cycle.

Table 6.3 Coleopterous phloem-boring insects found in Ghanaian forests

Species	Host plant	Life history/comments
Cerambycidae		
Acanthophorus spinicornis F.	Antiaris africana	Large longhorn beetle which usually breeds in dead trees, but has been observed boring into stressed trees in Nigeria; pupation in sapwood
Acmocera compressa F	Celtis mildbraedii, Albizia zygia, Piptadeniastrum africanum, Albizia ferruginea, Bosqueia angolensis, Pterygota macrocarpa, Triplochiton scleroxylon	
Acmocera conjux Thoms	Blighia sapida, B. angolensis, Hymenostegia afzelii, C. mildbraedii, Cleidion gabonicum, Nesogordonia papaverifera, A. zygia	
Acmocera olympiana Thoms	A. africana, C. gabonicum, H. afzelii, C. mildbraedii	
Acridoschema isidori Chevr.	B. angolensis, C. mildbraedii, C. gabonicum, H. afzelii, Cola spp., A. zygia, B. sapida, Conopharyngia sp.	
Aderpas lineolatus Chevr.	B. angolensis, C. mildbraedii, C. gabonicum, H. afzelii, Cola spp.	
Alloeme rubra Thoms.	P. africanum	
Analeptes trifasciata F.	Ceiba pentandra, Spondias mombin, Adansonia digitata, Sterculia setigera, Anona senegalensis, Anacardium occidentale, Sclerocarya birrea, Lannea nigritana, Pseudospondias macrocarpa	Adults feed by gnawing bark of small branches often girdling them; eggs laid above girdled stem; large black and orange beetle; has damaged Eucalyptus in Uganda and Nigeria
Ancylonotus tribulus F.	C. pentandra, T. scleroxylon	
Apoeme lugubris Oliv.	N. papaverifera	
Brachynarthon aeneipennis Brng.	C. mildbraedii	
Callanthemis gabonicus Thoms	A. zygia. P. africanum	
Callichroma afrum L.	C. mildbraedii	
Carinoclytus reichenowi Qued.	P. africanum	
Ceroplesis buttneri Kolbe	C. mildbraedii, N. papaverifera	
Coptops aedificator F.	A. zygia, A. africana, Baphia nitida, C. mildbraedii	

Cordylomera spinicornis (F)	*Entandrophragma cylindricum, Meliaceae, Khaya grandifoliola, Khaya ivorensis, Lovoa trichilioides*	Slender green beetle; widely distributed; can infest mature standing *K. grandifoliola*; eggs laid in bark of the host; larvae tunnel under bark until fully grown; adults emerge during dry season
Crossotus mossosus Duv.	*C. mildbraedii*	
Dere nigrita Gah	*A. africana*	
Dichostathes quadripunctatus Chevr.	*Baphia* spp., *Albizia adianthifolia, A. zygia, Cassia siamea*	
Eulitopus glabricollis Murr.	*Teclea verdoorniana*	
Exocentrus freyi Brena		
Falsovelleda congolensis Hintz	*Cola* sp.	
Glenea divergevittata Brng.	*Cola* sp.	
Glenea giraffe Dalm.	*Cola* sp. *P. macrocarpa*	
Glenea vigitiduomaculata Thoms.	*Spathodea campanulata*	
Graciella pulchella Klug	*A. africana*	
Idactus ellioti Gah	*Cussonia barteri*	
Lasiopezus sordidus Oliv.	*Antrocaryon micraster*	
Lasiopezus variegator F.	*Lannea welwitschii*	
Macrotoma serripes F.	*A. adianthifolia, C. mildbraedii*	
Monochamus centralis Duv.	*Cola millenii*	Larvae mine shallow galleries; pupal cell packed with wood fibers; 4–5 months life cycle; wide range of tree age and size attached
Monochamus griseoplagiatus (Thoms.)	*E. cylindricum*	Normally breeds in logs or dead trees; females cut oviposition niches; larvae bore beneath the bark; development requires 4 months
Monochamus ruspator F.	*E. cylindricum*	
Ocularia ashantica Brng.	*Cola caricifolia, Cola* sp.	
Ocularia subaschantica Brng.	*N. papaverifera*	
Ocularia transfersefasciata Brng.	*B. angolensis, C. gabonicum*	
Oeax collaris Jord.	*Ficus exasperate*	

(continued)

Table 6.3 (continued)

Species	Host plant	Life history/comments
Oeax lichenea Duv.	*B. angolensis, Lannea welwitschii*	Widely distributed in tropical Africa; brown long-horned borer; breeds in dead or fallen trees; larvae bore between bark and sapwood; pupation in wood
Olenecamptus trilagiatus Jord.	*C. gabonicum*	
Parandra gabonica Thoms	*Funtumia elastica*	
Phryneta leprosa F	*A. africana, Khaya* spp.	
Plocaederus fucatus Thoms	*Lannea welwitschii*	Living and recently felled trees are attacked; larvae tunnel beneath bark and pupate in sapwood
Plocaederus basalis Gah	*Guarea thompsonii, Terminalia ivorensis*	
Prosopocera bipunctata Drury	*Baphia pubescens*	
Pseudhammus occipitalis Lam	*A. zygia*	
Pterolophia burgeoni Brng	*P. macrocarpa*	
Rhopalizus nigripes Chevr	*P. africanum*	
Sternotomis amabilis Hope	*Ficus* sp.	
Sternotomis chrysopras Voet	*B. angolensis, Ficus* sp.	
Sternotomis mirabilis Drury	*B. angolensis, Ficus* sp.	
Sternotomis pulchra Drury	*A. africana, Milicia excelsa, Lannea welwitschii, Ficus* sp., *B. angolensis*	
Velleda callizona Chevr	*Trichilia heudelotii*	
Xylotrechus fragilis Jord	*Celtis adolfi, C. mildbraedii, N. papaverifera*	
Zoocosmius viridicinctus Auriv	*A. adianthifolia, A. zygia, P. africanum*	
Zoocosmius viridicinctus Auriv	*A. adianthifolia, A. zygia, P. africanum*	
Buprestidae		
Agrilis fraudulentus Per	*C. gabonicum*	

Species	Host	Biology
Agrilis hastulatus Fahrs.,	*A. adianthifolia, A. ferruginea, A. zygia*	Common species that oviposits in niches cut by females in bark of recently killed trees; oviposition may occur within 48 h of tree death; larval tunnel meanders between bark and sapwood; pupation in curved chamber in sapwood; develop from egg to adult in 8 weeks; adults fly in June, September, October
Agrilis nitidifrons Kerr	*Azelia africana*	Similar galleries and biology to *Agrilis hastulatus*, pupal chamber along wood grain; adults fly in July
Agrilus portonovensis Kerr	*A. adianthifolia*	Attack dead trees after they have dried; larvae tunnel between bark and wood; larvae bore into wood to pupate; adults bore a triangular curved exit hole; adults fly in February
Agrilus rohkirchi Obenb	*H. afzelii*	
Belionota canaliculata F	*T. scleroxylon*	
Chrysobothris dorsata F	*T. scleroxylon*	Oviposition on newly felled trees; larval tunnels between bark and wood; larval galleries 1–2 cm wide × 30 cm length; pupation within sapwood; exit hole oval; egg to adult development requires 4 months; adults fly July, August
Megactenodes ebenina Qued	*Chrysophyllum* sp.	
Megactenodes punctata Silb	*H. afzelii*	
Megactenodes westermanni Cast. Sp. And Gory	*B. nitida, C. mildbraedii*	
Spenoptera neglecta Klug	*B. angolensis*	

Figure 6.11 Powderpost beetle damage to wawa board (end view). Note the frass is lightly packed in the galleries

Figure 6.12 *Bostrychoplites cornatus* is a typical adult powderpost beetle (Bostrichidae). Note the sculptured pronotum with distinct spine and head pointed downward

Damage

Bostrichids most frequently breed in sawn and seasoned timber. Species of the genera *Xyloperthoides*, and *Heterobostrychus* are common pests of sawn lumber in West Africa. Other species, such as *Apate monachus* and *A. trebrans*, affect lumber but also attack green logs and weakened or dying trees. Bostrichids attack on living

Figure 6.13 Powderpost beetle damage to wawa used in crating high-value odum for export. (Ehwia Wood Products, Kumasi.)

trees, though important, is far less significant than damage to lumber and wood products. Damage in sawmills and lumber yards and to manufactured products results in tremendous economic losses (Figure 6.13). Since bostrichid larvae feed on wood starch, attack is restricted to the starch-rich areas of the tree, mainly the sapwood. The sapwood of all species of timber in Ghana is susceptible to attack. Those with high-starch contents such as wawa (*Triplochiton scleroxylon*) are extremely susceptible.

At times heavy attack will go unnoticed because the insects leave a skin of wood intact while destroying the interior. The fine powdery frass is the key to identifying attack by powderpost beetles.

Pest Management

Since Bostrichid attack is restricted to the sapwood, control can be achieved by eliminating sapwood from lumber and manufactured articles. Most sawmills in Ghana follow this practice when cutting high quality hardwoods. Foreign buyers of cut lumber often insist that wood must be "free of sap." In the case where lower grades of lumber are used, sapwood is often permissible. Some species, such as wawa, have no true heartwood. In these cases, removal of sapwood is impracticable and other control measures must be found.

Dipping or spraying sawn lumber with insecticides is a common chemical control for powderpost beetles (Plate 57). The most common insecticide historically used in West Africa is gBHC. In Ghana it is sold by the trade names Gammexane and Hexaplus. Generally, a 0.5–1.0% solution of gBHC used as a dip or spray is adequate. Wood preservatives like creosote are also effective in deterring attack of powderpost beetles.

None of these control techniques, however, eradicates insects which have already infested a piece of wood. In this case, burning the wood or heat sterilization are the only alternative. Burning destroys the wood, but is a very effective infestation control in a lumber yard. Heat sterilization is also very effective, but is more expensive and cannot prevent subsequent infestations. Heat sterilization followed by chemical treatment is an effective control measure.

Lyctidae

The second important family of powderpost beetles is the Lyctidae. This family is much smaller than Bostrichidae, containing only about 90 species. Similar to bostrichids, lyctids attack only the sapwood. However, one major difference is that lyctids attack only hardwood while bostrichids occur in both conifers and hardwoods.

Description and Life History

The insects are small, brown colored, and no more than 4 mm long. Their body is elongated and somewhat flattened. Unlike the Bostrichidae, lyctids do not possess a hood-like prothorax and the head is visible from above. The antennae are club-shaped with two joints (Jones 1959a, b).

The female deposits eggs in the pores of the sapwood (on the end grain). She does not bore into the wood, but rather deposits her eggs using a sharp ovipositor. The larvae hatch after 7–10 days and feed on the remnants of the egg shells. Second instar larvae begin feeding on the wood leaving a mass of tightly packed powdery frass in the tunnel. Pupation occurs near the surface of the wood. The emerged adults bore out of the wood. Mating occurs outside the wood and the female flies to infest other wood. The length of the life cycle varies with the wood starch content and other environmental factors.

Damage

Lyctidae infest seasoned sapwood of hardwood timbers and do not attack green timber. The appearance of attack is much like that of bostrichid, a fine powdering frass packed in the boring tunnels. Two Lyctid genera are by far the most important; *Lyctus* and *Minthea*.

In Ghana, *Minthea obsita* is a common pest of sawn lumber. In particular, *M. obsita* attacks sapele (*Entandophragma cylindricum*) and wawa (*T. scleroxylon*). *Lyctus brunneus* is a serious pest of plywood in Nigeria, but as of yet does not occur in Ghana.

Pest Management

Control measures for Lyctdae are identical to those for Bostrichidae. The important forest products pests of Ghanaian woods are summarized in Table 6.4.

Freshwater Wood Borer

Relatively few books written on the topic of forest insects would include a section on freshwater wood borers. That is because there are relatively few places in the world where this species becomes a problem. In Ghana and other parts of Africa, there is an interesting insect species that bores into wood underwater. It can cause considerable damage and is quite unique among wood borers.

Taxonomy and Biology

The insect responsible for damage to wood underwater is *Povilla adusta* Navas (Ephemeroptera; Polymitarcidae). This insect has been observed damaging wood underwater in Nzulezo, a village in the western region of Ghana, along Volta Lake, and in various locations in East Africa (Ampong 1977). Ephemeroptera (mayflies) exhibit gradual metamorphosis from egg through several nymphal instars to the adult. Mayflies are the only insects that undergo an additional molt as an adult. The nymphal stage bores into wood to serve as habitat but does not feed on the wood. The nymphs line the galleries with a silk-like material and feed on microscopic algae (Ampong 1977)

Damage

Povilla adusta larvae are aquatic and attack only occurs underwater. Therefore, this insect damages wood that is used as supports for fishing docks or house supports in locations like the village of Nzulezo where houses are built on stilts (Figure 6.14). Attacks are most concentrated at the mudline and boring is generally along the grain (Ampong 1977). Wood (even that of susceptible species) normally needs to be submerged for a year or so before attacks occurs. Damage occurring in freshwater and the silk-lined galleries are diagnostic for *P. adusta* damage.

Table 6.4 Insect pests of logs, lumber, and forest products in Ghana

Species	Host plant	Life history/comments
Powderpost Beetles (Coleoptera: Bostrichidae)		
Apate degener Murr	*Albizia zygia*	Relatively uncommon species
Apate monachus F	*Antiaris africana, Baphia nitida, Bosqueia angolensis, Celtis mildbraedii, Khaya grandifoliola, Morus mesozygia, Nesogordonia papaverifera, Dalbergia sissoo, Trichilia heudelotti, Azadirachta indica*	Black beetle whose larvae bore into felled trees; adults injure young trees by tunneling into small stems causing breakage; sapwood reduced to fine dust; oviposition tunnel oriented in the direction of wood fibers
Apate reflexa Lesne		Common species in Nigeria
Apate terebrans Pallens	*Cassia nodosa, Cedrela odorata, Dalbergia sissoo, Glyricidia maculate, Tectona grandis, Ceiba pentandra, Triplochiton scleroxylon, T. heudelotti, A. indica*	Throughout tropical Africa; large black beetle; adults bore into stems of young healthy trees; galleries along wood grain up to 12 in in length; tightly packed frass
Bostrychoplites cornutus Oliv	*Chlorophora excelsa, Khaya ivorensis, Terminalia superba, T. scleroxylon*	Eggs laid within short galleries; larvae tunnel longitudinal galleries in the sapwood; adults found at night; attacks sawn and seasoned timber
Bostrychoplites productus Imh	*Baphia pubescens, B. angolensis, N. papaverifera, Pachystela brevipes*	Collected from standing trees; important species; small diameter; attacks stem near base of the tree; vertical tunnel constructed upwards from entrance; adults attracted to light
Bostrychopsis tonsa	*Terminalia ivorensis*	Heavy attack on 1–2-year-old plantations; widely distributed in Africa; breeds in dead or fallen sapwood; causes wind breakage; adults attracted to lights at night
Dinoderus bifoveolatus Woll	*Milicia excelsa, T. superba, T. scleroxylon, C. pentandra, Oxytenanthera abyssinica, Pterygota macrocarpa*	Attacks *T. scleroxylon* within short period of felling; sapwood attracts beetles; tunnels separated by thin walls when populations high; egg–adult development requires 3 months
Dinoderus minutus F	*Albizia adianthifolia, Oxytenanthera abyssinica*	Breeding habits similar to *Donoderus bifoveolatus*
Heterobostrychus brunneus Murr.	*Entandrophragma cylindricum, T. scleroxylon*	Can attack freshly cut logs; common infestations in roofing and flooring timbers; attack brush-treated creosote lumber; important pest; prefer unbarked wood

Species	Host	Biology
Sinoxylon brazzai Lesne	*B. pubescens*, probably other hosts	Common in Nigeria; adults breed in felled moderate to large tree; entrance hole leads to mating chamber; paired circumferential galleries lead from mating chamber; adults attracted to lights at dusk in February–April
Cyrtogenius cribipennis Schedl	*T. scleroxylon*	Monogamous; irregular egg gallery; new species identified from Ashanti region, Ghana
Diamerus impar Chap.	*B. angolensis, Ficus* sp., *Ficus exasperate, T. scleroxylon*	Occurs in fire-killed trees in the southern savannah; general habits similar to *D. impar*
Hylesihopsis dubius Egg.	*B. angolensis, M. mesozygia, Guarea thompsonii*	Small common beetle in high forest zone of Ghana; gregarious attack pattern; monogamous; single egg galley along grain of bark, larval galleries radiate; adult flight diurnal
Hypothenemus cassavaensis		Adults fly in June
Hypothenemus polyphagus (Egg.)	*Mangifera indica, Cassia siamea, T. scleroxylon*	Quite variable species in high forest; usually found in twigs and small branches; also feeds in fruits and seeds; can attack living twigs
Pityophthorus busseae Schedl	*Bussea occidentalis*	Moderate polygamy; nest is stellate; long egg galleries occasionally branched
Pityophthorus joveri		
Polygraphus granulatus Egg	*Carapa procera, Cedrela odorata, Khaya anthotheca, K. ivorensis*	Breeds only in cut trees and prefers branchwood; adults attracted to lights at night; adults fly in April
Polygraphus granulifer Egg	*Meliaceae*	Prefers dead trees and large branchwood; adults common at lights; adults fly in April, July, August
Xylion securifer Lesne	*K. ivorensis, A. zygia, Piptadeniastrum africanum*	Attacks felled trees; entrance with small mating chamber; circumferential oviposition tunnel along wood grain; larvae bore shallow tunnel; pupation at end of larval gallery
Xylopertha crinitarsis Imh	*M. excelsa, Mitragyna stipulosa, K. ivorensis, A. adianthifolia, A. zygia, B. nitida, B. pubescens, B. angolensis, C. mildbraedii, Pachystela brevipes, P. africanum, T. heudelotti, T. scleroxylon*	Common and important pest in recently felled trees; entrance tunnel expands to mating chamber; circumferential oviposition tunnel on either side of entrance approximately 6 in; tunnels tightly packaged with boring dust

(continued)

Table 6.4 (continued)

Species	Host plant	Life history/comments
Xylopertha picea Oliv	*M. excelsa, A. africana, T. scleroxylon*	Common species in Takoradi harbor; often attacks stems girdled by *Analeptes trifasciata*: oviposition tunnel is excavated in the direction of the wood fibers
Xyloperthoides nitidepennis Murr	*M. excelsa, T. scleroxylon, A. adianthifolia, A. zygia, B. pubescens*	Attacks recently felled trees; entrance tunnel with mating chamber; oviposition tunnels always horizontal in standing tree; larvae bore in direction of wood fibers; development egg–adult 3 months; diurnal adult flight
Xyloperthoides orthogonius Lesne	Polyphagous	Common species in savannah; attacks small stems and branches of recently felled trees; adults emerge late afternoon
Powderpost Beetles (Coleoptera: Scolytidae)		
Chortastus similes Egg	*Musanga cecropioides*	Monogamous bark beetle that generally attacks large branches; main nest is irregular pairing chamber and transverse egg gallery; larval galleries follow grain of bark
Ctonoxylon bosquieae Schedl	*B. angolensis*	Common small bark beetle; monogamous; nest with short egg gallery running across the grain with nuptial chamber; larval tunnels irregular along the grain
Ctonoxylon flavescens	*T. scleroxylon*	Common species throughout forest zone of Ghana; adults fly in January, April

Figure 6.14 Wood damaged by the freshwater wood borer *Povilla adusta* from Nzulezo in western Ghana. (Photograph courtesy of F.K. Ampong, FPRI, Kumasi.)

Pest Management

The most appropriate and effective means of controlling the insect is the use of resistant wood species. Ampong (1977) determined that *Milicia excelsa, Nauclea diderrichii*, and *Chrysobolanus ellipticus* are quite resistant. Copper-chromated-arsenic (CCA) is not an effective wood treatment and should not be used in aquatic situations. Cobbinah (unpublished) observed that coating wood surfaces with oil paint prior to use in aquatic medium reduces incidence of attack.

Miscellaneous Wood Borers

There are obviously many other species of insects that are known wood borers. Examples include *Zographus regalis* (Figure 6.15), *Sternotomis pubchabifactia* (Figure 6.16), *Cloniophorus chrysaspis* (Figure 6.17), *Phosphorus virescens* (Figure 6.18), *Petrognatha gigas* (Figure 6.19), *Megactenodes punctuator* (Figure 6.20) and Xylocopidae (Figure 6.21).

Figure 6.15 *Zographus regalis*, a cerambycid wood borer of wawa (*Triplochiton screloxylon*) (2.4 cm body length)

Figure 6.16 *Sternotomis pubchabifactia* wood borer of *Antiaris africana* (2.4 cm body length)

Figure 6.17 *Cloniophorus chrysaspis* wood borer of *Milicia excelsa* (2 cm body length)

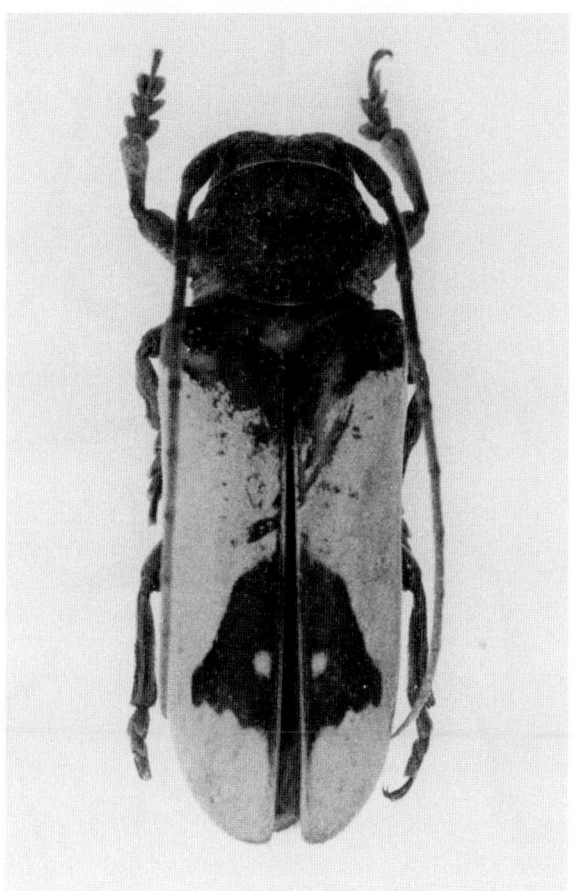

Figure 6.18 *Phosphorus virescens* wood borer of *Sterculia tragacantha* (3.5 cm body length)

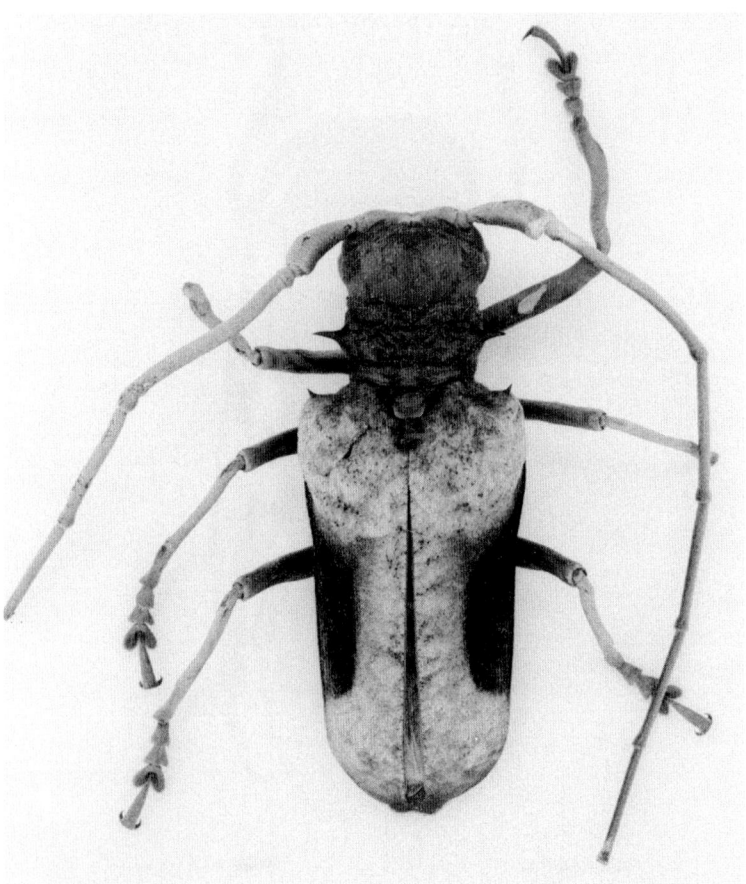

Figure 6.19 *Petrognatha gigas* wood borer of *Celtis zenkeri* (6.3 cm body length)

Figure 6.20 *Megactenodes punctuator* wood borer of *Hymenostegia afzelii* (2.4 cm body length)

Figure 6.21 Wood-boring carpenter bee (Xylocopidae) from Ghana; (host – *Ficus* sp.)

Chapter 7
Termites

Introduction

Termites are a common and important insect order in the warmer regions of the world, including West Africa. Their presence can hardly go unnoticed to inhabitants or travelers of tropical countries. In some areas of Africa, termite mounds dominate the landscape (Plate 58). Since termite food is chiefly wood and woody tissues, these insect cause many problems to the users of wood. Telephone poles, wooden beams, and animals' pens are often rendered useless by termite attack. Live trees and plants can also be damaged by termite activity. Subterranean termites create mats of woody tissue within their mounds upon which they cultivate fungi that is a significant food source (Plates 59 and 60).

Termite populations can reach very high numbers but, because of their cryptic habits, termites are often unnoticed by the average citizen. Only during the early part of the rainy season do they become obvious. Termites spend a relatively brief period of time in the winged stage (alate) away from the nest. When the alate leaves the nest it is only for a purpose of starting new colonies. Large swarms (sometimes referred to as "rain flies") can be seen after heavy rains. Many species are crepuscular or nocturnal and attracted to light at night. It is only during the mating flights that termites lose their normal cryptobiotic habit. The cryptic habit of termites belies their economic and ecological importance.

Termite Castes

Like other social insects, individual members of termite colonies are divided into castes; each caste carries out specific responsibilities. Castes of termite include the primary reproductive, the secondary reproductive, and the sterile soldiers and workers. Occasionally major and minor workers as well as major and minor soldiers occur. The primary reproductives are the winged forms and they are to be regarded as the ancestral caste from which all other forms including the king and queen have been derived. Pairs of adult winged reproductives (alates) after shedding their wings and

Figure 7.1 Typical termite queen (11 cm body length). In the foreground is the king termite (1 cm body length)

mating may succeed in founding a new colony where the pair will become king and queen. The king's appearance remains similar to the alate's but the queen will gradually develop a very large abdomen over a period of years as capacity to produce eggs increase. Queen termites 8–10 cm in length are common among the family Termitidae (Figure 7.1, Plate 61). The number of individuals in various castes is a function of the age of the colony and other environmental conditions.

The worker termite is a nymph whose development has stopped during the second instar. Workers are similar to adults but possess more lightly chitinized heads. The main function of the worker is to provide food for the colony and build the nest. The soldier termite is much like the worker except that the head has greatly developed mandibles. Soldier termites usually rush to any damaged portion of a termite mound to defend against intruders. Soldiers of many species are capable of inflicting serious bites on humans.

Caste Determination

Caste determination is one of those areas of basic insect biology that has attracted considerable research attention. It has been shown that nymphs removed from a colony and kept under suitable environmental conditions will develop into all castes (Harris 1961). When large numbers of one caste are removed from a nest, nymphal development of that caste increases until the castes are in a balanced condition. This balancing phenomenon is thought to be a result of the close contact between all individuals through reciprocal grooming, rubbing on antennae, and exchange of food. During these contacts the number of contacts with individuals of a given caste is in a sense "tallied." This tally is sent to the queen via chemical signals. The queen can then use this information to produce chemical signals that circulate among all the caste and regulate the social behavior of the colony.

Colony Formation

According to Harris (1961), new termite colonies can be formed in one of three ways: (1) pairs of winged reproductives that leave an existing nest and establish a new colony; (2) isolation of part of an existing colony that has secondary reproductives; and (3) migration of part of the colony.

The most common method of colony formation is via winged reproductives or alates. At specific times of the year, in Ghana usually at the beginning of the main rainy season, large numbers of alates leave existing mounds. It is at this time that the huge populations of termites become obvious. The alates usually leave the existing nest through special openings created by workers just for the occasion. Most species leave the nest in one or a few nights. When a large number of termites exist in an area the populations of alates emerging at one time is huge. The morning following alate flight thousands of dead termites and their cast wings can be found beneath night lights. Sometimes these will accumulate to several centimeters in depth. Few alates will actually successfully establish new colonies, but with such large numbers there is little danger of the termite species becoming extinct.

Alate flights are usually short and may last only a few hours. After flights the alates lose their wings and begin searching for a mate. On encountering a dealate of the opposite sex, they show some mating behavior, for instance, the female can be observed raising her abdomen in an attractive manner to entice the male. Harris (1961) has suggested that scent signals are produced by the female. The male then follows the female in tandem fashion until a suitable nesting place is found. Mating usually occurs after the royal cell has been completed. After mate selection the adults quickly lose their attraction to light and seek darkness and shelter. When a suitable shelter is found the pair begins a new colony. Initially, the biggest problem facing the new colony is available food since the adults do not leave their cell. The first few eggs give rise to workers who begin the job of feeding the colony. A few soldiers may also develop from the first eggs laid. When enough workers are available for food gathering and nest construction, the adult pair becomes exclusively concerned with egg production. The egg production of the queen increases sharply as the colony develops. In some species the mature female can lay an egg every 2 seconds (Harris 1955).

Eventually the original parents reach the end of their reproductive capacity, or are unable to keep up with the reproductive demands of the colony. In this case, secondary reproductives develop. Up to 200 of these auxiliary egg producers may exist in the colony, depending on the species; their main function is egg production. Individually, secondary reproductives do not produce as many eggs as the queen, but collectively produce far more. These secondary egg layers can maintain the colony in the event of the queen's death, which is why removing a queen from a termite colony is not a guarantee that the colony will die.

The formation of a new colony by isolation depends on the successful development of secondary reproductives. When the original nest is very large "budded" colonies can form. Budded colonies have the advantage that a full complement of all castes are present, avoiding the very vulnerable first phase of colony formation by winged adults. Budding does not occur in all species of termites and is far less important than colony formation by winged reproductives. Colony formation by migration is a relatively rare event and has only been observed for a few species (Harris 1961). Events that initiate this behavior are poorly known, but all castes have been observed leaving a nest and establishing an entirely new nest nearby. This colony-formation method is restricted to higher termites that are capable of producing secondary reproductives.

Identification

Termites constitute the insect order Isoptera. This order is represented by six families and about 2,000 species. The six families are: Mastotermitidae, Kalotermitidae, Termopsidae, Hodotermitidae, Rhinotermitidae, and Termitidae. Common characteristics of termites are biting mouthparts, filiform antennae, and two pairs of similar membraneous wings longer than the body. The Isoptera comes from the root *iso* meaning "equal" and *ptera* meaning "wings." Metamorphosis is incomplete and all species are social insects. The taxonomy of termites is quite stable.

The identification of termites to the species is a specialized task generally beyond the capability of the practicing forest or forest pest management specialist. However, termites can be keyed to the family with relative ease using the keys provided (Tables 7.1–7.3).

Table 7.1 Key to families of alate termites

1. Tarsi distinctly 5 segmented with pulvillus; antennae with about 30 segments; hindwing anal lobe ....Mastotermitidae
1. Tarsi 4-segmented, viewed from above; antennae rarely with more than 27 segments; hindwing without anal lobe..2
2. Anterior wing scales large enough to cover the posterior scales; wings reticulate 3
2. Anterior wing scales short; not reaching to base of posterior scales; wings not wholly reticulate ....Termitidae
3. Ocelli; present 4
3. Ocelli; absent 5
4. Fontanelle present..Rhinotermitidae
4. Fontanelle absent ....Kalotermitidae
5. Pronotum saddle-shaped; tarsi definitely 4-segmented ....Hodotermitidae
5. Pronotum flat; tarsi viewed from below seen to posses a rudimentary 5th segment..Termopsidae

Table 7.2 Key to the families of workers termites

1. Right mandible with a subsidiary tooth present at the base of the first marginal tooth
.... Hodotermitidae and Termopsidae
1. Right mandible without a subsidiary tooth at base of first marginal tooth
.... 2
2. Left mandible with a terminal group of 3 teeth (apical, first and second marginal)
.... Rhinotermitidae
2. Left mandible with at most 2 teeth in terminal group (apical and first marginal)
....Termitidae

Note: A true worker caste is considered absent in the Mastotermitidae and Kalotermitidae; the older nymphs can be determined from the key to the alate characters.

Table 7.3 Key to the families of soldier termites

1. Tarsi 5-segmented ....Mastotermitidae
1. Tarsi 4 segmented, rarely with rudimentary fifth segment.......................................2
2. Pigmented eyes and abdominal cerci present...3
2. Pigmented eyes and abdominal cerci absent 3
3. Head rounded, generally subconical..Hodotermitidae
3. Head flattened, more angular... Termopsidae
4. Fontanelle present..5
4. Fontanelle absent ....Kalotermitidae
5. Pronotum flat without anterior lobes...Rhinotermitidae
5. Pronotum saddle-shaped with anterior lobes..Termitidae

Note: The soldier caste is absent in the genera *Anoplotemmes* and *Speculitermes*.

Mastotermitidae

This family is represented by a single species occurring in northern Australia. The eggs are laid in groups while all other termites lay eggs singly.

Kalotermitidae

These are the drywood termites, so named because they live in wood with no contact with the ground or soil. Their only source of water is the moisture in the wood. They are common insects throughout the tropics and subtropics of the world. The worker caste is absent.

Termopsidae

Termopsidae are the dampwood termites. Structurally they are relatively primitive and usually found in subtropical and temperate zones. Colonies tend to be small and the worker class is absent. Their habitat is damp wood.

Hodotermitidae

This family forages for food and is known as harvester termites. All castes have functional compound eyes and two soldier castes are present.

Termitidae

This family includes 80% of all termite species. They are mostly wood-eating and either subterranean or mound-builders; a few species build arboreal nests. The termites that cultivate fungus gardens that are very common in Ghana occur in the subfamily Macrotermitinae.

Ecological Groups of Termites

Harris (1955) divided the termites into four ecological groups: (1) drywood termites; (2) dampwood termites; (3) subterranean termites; and (4) ground-dwelling termites. For purposes of this book we will discuss drywood termites, subterranean termites (including subterranean and ground dwelling of Harris (1955), and termites attacking living trees. Dampwood termites are a small group of termites that are of limited importance locally.

Drywood Termites

Drywood termites are in the family Kalotermitidae. *Cryptotermes havilandi* is the most common species in Ghana and occurs abundantly along the coast, but also inward to the Ashanti region (Williams 1973) (Figure 7.2). *Cryptotermes brevis* also occurs in Ghana, but is much less common (Williams 1973). Drywood termites are easily moved around the world in wooden furniture and shipping crates and their introduction to new areas is highly probable.

Description and Life History

The winged adults (alates) of drywood termites are similar to most other termite species. They have long wings that extend well beyond the body. The alates can be found at most times in colonies established in sound dry wood, which is unlike the situation with subterranean termites. The adults are smaller than subterranean termites, rarely exceeding 6.3 mm in length.

Soldiers represent a small proportion of the colony population, but are easily recognized by the flattened forehead that is dark in color. Nymphs form the bulk of the colony and produce characteristic hexagonal frass pellets. The worker caste is absent in this family.

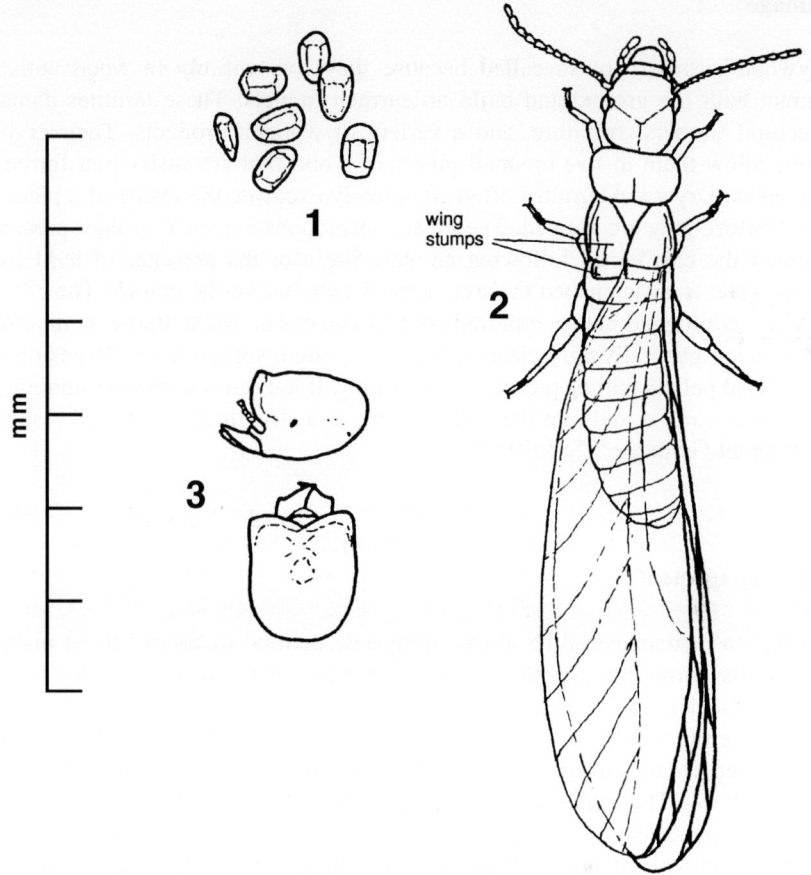

mm

wing
stumps

1

2

3

Figure 7.2 *Cryptotermes havilandi* is a common drywood termite species along the Ghana coast: (1) frass; (2) adult; (3) head. (From Anonymous 1970; reprinted with permission from Buildings and Roads Research Institute, Kumasi.)

Colony formation begins with the alate mating flight. Following the nocturnal flight, the alates lose their wings and form pairs. The male and female pairs burrow into wood, closing the entrance hole with wood and saliva. The first nymphs are fed by the king and queen, but later nymphs undertake the work of the colony. Activities of the nymphs include excavating the gallery, digesting wood, and feeding the royal pair, soldiers, and younger nymphs. There is no special queen's chamber, as occurs in the nest of some of the subterranean species, so she is able to move freely around the nest. The queen lays a few eggs at a time and never reaches the size of the subterranean queens. The average drywood termite colony would include only a few hundred individuals, but a very successful colony of *C. havilandi* may develop to 5,000 individuals within 5 years (Anonymous 1970).

Damage

Drywood termites are so called because they live entirely in wood with no contact with the ground and build no earthen tunnels. These termites damage structural timbers, furniture, and a variety of wooden products. Their cryptic habits allow them to live in small pieces of wood that are easily transferred to new areas. Drywood termites often completely excavate the inside of a piece of wood before they are noticed. The most conspicuous evidence of their presence is either the cast wings following an alate flight or the presence of hard frass pellets. The frass is ejected through a small entrance to the colony. The colony entrance is plugged between periods of frass ejection. The galleries of drywood termites are generally very clean and do not contain soil particles. Occasionally loose fecal pellets will be present in the colony. If the infestation goes unnoticed, the colony can expand into the entire length of a structural timber and result in its complete collapse.

Pest Management

Prevention of damage is the most appropriate method to control these insects. The adults normally enter through the end grain of wood or through joints in wood construction. Treating all exposed ends and ensuring tight wooden joints can reduce the incidence of these pests. The best method is to chemically treat all exposed wood with a good wood preservative. Oil-based paints also deter termite attack. The impregnation characteristics of the wood may be important to ensure that adequate preservative is present to deter these pests. Untreated wood should never be used in housing construction. Probably the most effective preventative control is the use of resistant species of wood. This method of prevention has been advocated by Quao (1965), Harris (1961), and Anonymous (1970). Table 7.4 indicates the relative resistance characteristics of some Ghanaian wood species.

Remedial control is relatively more difficult than prevention. Though drywood termites are susceptible to a variety of chemical insecticides, getting the chemical in contact with the insects is difficult. Valuable buildings can be fumigated with methyl bromide or other available fumigants. This method will kill the insects currently active, but will not prevent reattack later. Insecticide can be injected into the wood through small holes drilled the length of the infestation. Quao (1965) recommends 12.7 mm (1/2 in) holes drilled every 61 cm (2 ft) along the damaged member. Chemicals that have some degree of persistence are recommended. A final remedial approach is to replace all the damaged material with resistant or chemically treated wood. This method will likely provide the most lasting control.

Table 7.4 The relative resistance of some of the major Ghanaian wood species to attack by *Cryptotermes havilandi*, a common drywood termite. (Modified from Anonymous 1970.)

Susceptible	Slight resistance	Moderate resistance	High resistance
Amphimas pterocar-poides	*Bosquiea angolensis*	*Afzelia africana*	*Albizia sp.*
	Entandrophragma candollei	*Milicia excelsa*	*Distemonanthus benthamianus*
Anopyxis calaensis		*Cylicodiscus gabon-ensis*	
Antiaris africana	*Lovoa klaineana*		*Erythrophleum guineense*
Berlinia sp.	*Piptadeniastrum africanum*	*Entandrophragma cylindricum*	
Daniellia ogea			*Mimusops heckelii*
Entandrophragma angolense	*Terminalia ivorensis*	*Pterocarpus scyauxii*	*Sarcocephalus diderrichii*
		Terminalia superba	
Guarea thompsonii			*Staudtia stipitata*
Khaya ivorensis			
Mitragyna ciliata			
Triplochiton scler-oxylon			

Subterranean Termites

Members of this group are made up of three families of termites: Hodotermitidae, Rhinotermitidae, and the Termitidae. Because this group encompasses approximately 80–90% of all the termites in a given locality, it is difficult to assign a simple taxonomic basis for identifying these termites. However, identification by habitat is simple; any termite that lives in the soil or in direct contact with the soil and any termite in a mound is a member of this group.

Description and Life History

Many of the general discussions on termite castes and colony formation discussed at the beginning of this chapter refer to this group of termites. Members of this group build a central nest (either in the soil or in contact with the soil) and move out from the nest in search of food. Unlike drywood termites, subterranean termites frequently occur in very large numbers; populations may reach 2–3 million individuals in a large termite mound (Harris 1955) (Plate 58). It is within these large termites nests that the much enlarged queens can be found, usually completely enclosed in a hardened clay chamber called the "royal chamber" (Plate 61).

Associated fungi play an important role in the ecology of termites. Fungi may be important in breaking wood down prior to termite attack, or may be cultivated by termites to breakdown wood or as a food source. Usher (1978) has discussed the role of fungal deterioration of wood in making wood suitable for termites. Some termite species, such as *Ancistrotermes* spp. and *Pseudacanthotermes militaris*, require some fungal deterioration before they will attack wood, while *Macrotermes* spp. and *Microtermes subhyalinus* can exploit wood with relatively little fungal damage (Usher 1978). Other groups of termites cultivate fungus gardens within their mounds (Figure 7.5). These fungus gardens are cultivated on convoluted mats of woody tissue brought into the termite mound (Figure 7.3, Plate 60). These termites apparently cultivate and feed on the fungus directly rather than consuming the wood. Periodically the woody mats are dismantled and moved out of the nest to permit fruiting of the fungus. In other cases a fruiting body arises directly from the mats in the soil. One species of fungus in Ghana, *Termitomyces* spp., develops very long and tough fruiting stalks

Figure 7.3 Convoluted mat of woody tissue with evidence of fungus taken from a subterranean termite nest near Kumasi

from mats in the soil. These fruiting bodies can reach 0.6 m in length (Lawson 1986) and are commonly eaten by humans in areas where they grow (Plate 62). These mats of woody tissue and fungi also help regulate humidity within the termite mound.

Damage

Because of the large nest size and dominance of this group of termites, subterranean termites probably account for 95% of all termite damage (Quao 1965). Wood destruction can occur quickly; a wawa board placed on the soil can be consumed by termites to the point of complete loss of structural strength in as little as 1 month.

These termites have developed a unique way of maintaining their "soil environment" while some distance away from the soil. Earthen tunnels are built by these termites, in some cases for great distances (Figures 7.4 and 7.5). These tunnels extend from the nest to distant sources of woody material and can traverse steel, concrete, or resistant wood to get to susceptible material (Figure 7.6). When woody material is encountered, the termites enclose it in a thin shell of soil to maintain a suitable environment while the wood is being consumed (Figure 7.7). These earthen tunnels and soil within the galleries are diagnostic of subterranean termite attack.

Pest Management

Control of termites, including subterranean termites in Ghana, has been reviewed by Quao (1965) and Mensa-Bonsu (1973). In general, control strategies are divided into preventive and treatment methods.

Prevention is the far superior approach and is often less expensive than post infestation treatment methods. Prevention includes proper preparation of the building site, preventative housing design, and use of resistant materials. Prior to construction, building sites should be cleared of all residual woody debris and graded to insure good drainage. In this way, termites are deprived of food sources near the building and available water is reduced. Areas under concrete foundations and floors should be sprayed with a persistent insecticide such as Aldrex 40 (Aldrin). Concrete floors should be poured in such a way as to minimize the likelihood of cracking, which will provide an entrance for termites.

Termite shields can also be an effective preventative approach. These are usually metal strips that are placed between the concrete and susceptible wooden members. Most shields will not actually prevent termites from crossing them, but rather expose termite tunnels making it easy to detect and destroy the tunnels as they appear. Also, houses that are elevated above the soil surface allow for easy inspection and removal of termite runways.

The use of termite-resistant wood, in addition to the above preventative measures, can greatly reduce the damage from termites. In Ghana, most buildings are

Figure 7.4 Termite earthen tunnel across a treated wawa board. The narrow tunnel often indicates this is a transport tunnel

Figure 7.5 Earthen tunnel of subterranean termite. The broad meandering tunnel often indicates the termites are feeding on the wood

Figure 7.6 Termite earthen tunnels allow these insects to traverse resistant material; in this photo, tunnels allow the termites to pass over the resistant living wood in an attempt to get to dead branches in the crown

constructed with relatively expensive concrete block while less expensive woods are available. Several aspects of resistance are important to effectively use termite resistant wood species. First, the location within a tree affects the resistance characteristics of the wood. In general, heartwood is more resistant than sapwood. But in a few species, such as *Anopyxis klaineana, Diospyros sanza-minika*, and *Klainedoxa gabonensis*, the sapwood is more resistant than the heartwood (Ocloo and Usher 1980). Detailed descriptions of the resistance characteristics of many Ghanaian woods are discussed by Usher and Ocloo (1979) and their relative resistance is presented in Tables 7.5 and 7.6. Unfortunately, only a few of the major economic species in Ghana are highly resistant to termite attack.

Figure 7.7 Wooden frame left outdoors that is completely covered with earth while the wood is being consumed within

Manufacturing processes can alter the resistance characteristics of woods. For example, wawa, *Triplochiton scleroxylon*, which is a susceptible species in solid form, has improved resistance when converted to a woodwool (Atuahene 1972). Also, chemical treatment with Permethrin, Dieldrin, Lindane and other compounds provide good protection for wood against subterranean termites (Ocloo 1983).

One final aspect of termite resistance to consider is that resistance is also a function of the age of the wood. After wood has aged for some years the resistance properties are leached or lost from the wood. Consequently, even very resistant wood will succumb to attack when under severe conditions for long periods of time. The use of resistant wood, good building design, and maintenance is all that is necessary to utilize wood in building construction in even the most severely termite infested area.

When termites have already infested a structure, control becomes more difficult. If termites have established a nest below the floor or foundation, holes are drilled through the floor or foundation and insecticide injected into the nest or nearby soil. Likewise, termite infestation in structural wooden members can be controlled by drilling into the wood and injecting chemical into the wood, similar to the methods suggested for drywood termites. Resistant wood should also be used when replacing damaged wooden members. When wood is attacked via the construction of earthen tunnels, merely breaking the tunnel repeatedly will prevent further attack and eventually kill any termites isolated in wood away from the nest.

Table 7.5 Relative resistance of economic species of Ghanaian woods to subterranean termites. (Modified from Ocloo and Usher 1980.)

Very resistant	Resistant	Moderately resistant	Susceptible
Lophira alata	Pericopsis elata	Entandrophragma angolense	Antiaris africana
	Milicia excelsa		Entandrophragma candollei
	Nauclea diderrichii	Entandrophragma cylindricum	
	Piptadeniastrum africanum		Guarea cedrata
		Entandrophragma utile	Guarea thompsonii
	Terminalia ivorensis	Tieghemella heckelli	Guibourtia ehie
			Khaya anthotheca
			Khaya grandifoliola
			Khaya ivorensis
			Lovoa trichilioides
			Mansonia altissima
			Nesogordonia papaverifera
			Tarrietia utilis
			Triplochiton scleroxylon
			Turraenthus africanus

Termites Attacking Living Trees

Several species of termites, including *Macrotermes bellicosus* (Smeath), *Macrotermes natalensis* (Hau.), *Coptotermes sjostedti* Holmgr, and *P. militaris* (Hag), have been reported as general pests of living trees throughout West Africa. *Ancistrotermes cavithorax* and *Amitermes evuncifer* are two termite species that limit the establishment of eucalyptus in drier areas of Ghana (Kudler 1970a–c). All of these species cause damage by attacking the main stem of the tree at or near the soil line and girdle the young tree (Figure 7.8 and 7.9).

Damage

Termite attack of living trees in Ghana is a potentially important problem facing the use of exotic forest species. Termite attack on native species tends to be limited to trees that are physically damaged, allowing entry to dead heartwood or overmature trees (Figure 7.10). When termites attack and kill young, fast growing exotic plantations, they can have important economic effects on afforestation and reforestation efforts.

Table 7.6 Relative resistance of secondary species of Ghanaian woods to subterranean termites. (Modified from Ocloo and Usher 1980.)

Very resistant	Resistant	Moderately resistant	Susceptible
Borassus aethiopum	Afzelia africana		Albizia adianthifolia
Cylicodiscus gabunensis	Albizia ferruginea		Albizia zygia
	Blighia sapida		Alstonia boonei
Cynometra ananta	Bussea occidentalis		Amphimas pterocarpoides
Gluema ivorensis	Combretodendron macro-carpum		
Khaya senegalensis			Aningeria robusta
Lophira lanceolata	Dialium aubrevillei		Anogeissus leiocarpus
Manilkara multinervis	Distemonanthus bentha-mianus		Anopyxis klaineanu
			Antrocaryon micraster
Manikara obovata	Mammea africana		Bombax brevicuspe
Pentaclethra macrophylla	Morus mesozygia		Bombax buonopozense
	Ongokea gore		Canarium schweinfurthii
Pseudocedrela kotschyi	Pterocarpus erinaceus		Carapa procera
	Sacoglottis gabunensis		Ceiba pentandra
	Sterculia rhinopetala		Celtis adolfi-friderici
			Celtis mildbraedii
			Celtis zenkeri
			Chrysophyllum albidum
			Chrysophyllum perpulchrum
			Cola gigantea
			Cola lateritia
			Daniellia ogea
			Daniellia oliveri
			Diospyros sanza-minika
			Elaeis guineensis
			Erythrophleum ivorensis
			Ficus capensis
			Holarrhenag floribunda
			Klainedoxa gabonensis
			Mitragyna stipulosa
			Musanga cecropioides
			Pterygota macrocarpa
			Strombosia glaucescens
			Terminalia superba
			Trichilia prieuriana

Pest Management

Control of termite attack on living trees is best accomplished by chemical and cultural methods. Susceptible tree species in nurseries can be treated using a drench solution of Dieldrin and Aldrin. Treating soil prior to planting seeds is also effective using these

Figure 7.8 *Ancistrotermes* girdling neem used in agroforestry project in Kumasi

persistent chemicals. In plantations, application of an Aldrin (Aldrex 40) drench around the base of attacked trees is an effective control (Kudler 1970a–c).

Cultural practices of avoiding exotics known to be susceptible to termites will minimize the likelihood of damage. There is some preliminary evidence that susceptibility to termite attack in eucalyptus is dependent on the species (Table 7.7). In general, exotic species should be tested for their susceptibility to termites prior to any wide-scale planting.

Beneficial Role of Termites

Although serious pests to wooden structures and living trees, termites also have beneficial roles on the tropical forest ecosystem. One of the most important roles is the cycling of nutrients. Termites consume roughly 20% by weight of the litter

Figure 7.9 *Caliandra* sp., an exotic tree species from east Africa, killed by girdling from *Ancistrotermes* sp. in an agroforestry project at Kwame Nkrumah University of Science and Technology, Kumasi

fall and 8–10% of annual primary productivity in the high forest of Ghana (Usher 1978). Because termites usually consume only dead wood, nutrients are recycled quickly in an ecosystem where nutrients are quite limiting. Termites also aerate the soil and improve water penetration. In some parts of Africa, termites play a role in the establishment of some tree species such as *Khaya ivorensis* (Harris 1955). Also, certain termite mounds are generally indicative of general soil conditions (Harris 1955).

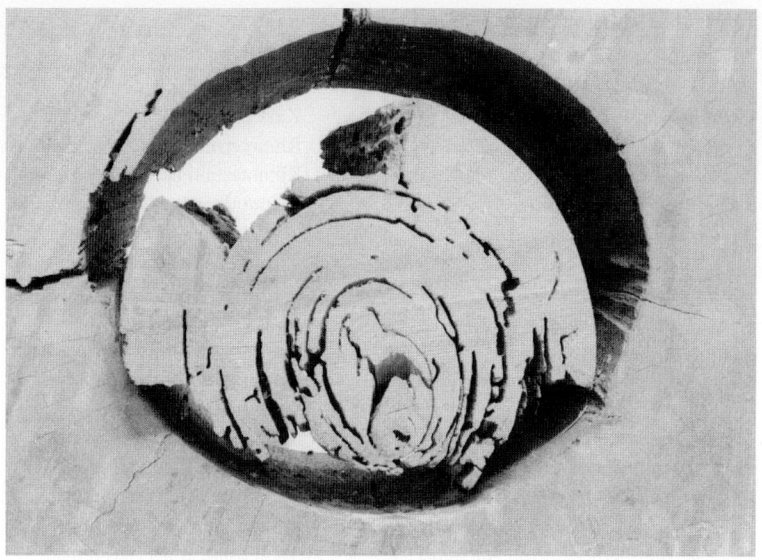

Figure 7.10 *Coptotermes* caused damage to the center heartwood of a living *Khaya* sp.

Table 7.7 Relative resistance of 12 species of living *Eucalyptus* to attack by the subterranean termites in coastal and Sudan savannah areas of Ghana[1]. (Adapted from Kudler 1970a–c.)

Moderate resistance	Intermediate resistance	Susceptible
E. tereticornis	*E. microtheca*	*E. torrelliana*
E. citriodora	*E. camaldulensis*	*E. grandis*
	E. cadambae	*E. salinga*
	E. coolobah	*E. bicolor*
	E. alba	
	E. populnea	

[1] Relative resistance based on several trials in different locations

Termites also play an important role in the food chain of African forests. Because of the abundance, termites are a major food source for other insects and vertebrates. In parts of Africa native peoples frequently consume termites either raw or cooked. They are delicious when properly prepared and serve as another excellent source of protein that can significantly improve the nutritional status of native diets. Termites are also an excellent source of protein for domestic poultry. Table 7.8 lists the wood-feeding termite species recorded in Ghana.

Table 7.8   Wood-feeding termites recorded in Ghana

Species	Family/subfamily
Cryptotermes havilandi (Sjostedt)	Kalotermitidae
Cryptotermes brevis (Walker)	Kalotermitidae
Coptotermes intermedius (Silvestri)	Rhinotermitidae: Coptotermitinae
Coptotermes sjostedti (Holmgren)	Rhinotermitidae: Coptotermitinae
Amitermes evuncifer (Silvestri)	Termitidae: Amitermitinae
Amitermes stephensoni (Harris)	Termitidae: Amitermitinae
Amitermes crucifer	Termitidae: Amitermitinae
Amitermes spinifer (Silvestri)	Termitidae: Amitermitinae
Anenteotermes polyscolus (Sands)	Termitidae: Amitermitinae
Astalotermes quietus (Silvestri)	Termitidae: Amitermitinae
Microcerotermes brachygnatus (Silvestri)	Termitidae: Amitermitinae
	Termitidae: Amitermitinae
Microcerotermes parvalus (Sjostedt)	Termitidae: Macrotermitinae
Ancistrotermes cavithorax (Sjostedt)	Termitidae: Macrotermitinae
Ancistrotermes crucifer (Sjostedt)	Termitidae: Macrotermitinae
Ancistrotermes guineensis (Silvestri)	Termitidae: Macrotermitinae
Macrotermes bellicosus (Smeathman)	Termitidae: Macrotermitinae
	Termitidae: Macrotermitinae
Macrotermes subhyalinus (Rambur)	Termitidae: Macrotermitinae
Microtermes subhyalinus Silvestri	Termitidae: Macrotermitinae
Odontotermes fidens (Sjostedt)	Termitidae: Macrotermitinae
Odontotermes pauperans (Silvestri)	Nasutitermitinae
Pseudacanthotermes militaris (Hagen)	Nasutitermitinae
	Termitinae
Nasutitermes latifrons (Sjostedt)	Termitinae
Nasutitermes lujae (Wasmann)	Termitinae
Allognathotermes hypogeus (Silvestri)	Termitinae
	Termitinae
Basidentitermes mactus (Sjostedt)	Termitinae
Cubitermes subcrenulatus (Silvestri)	
Procubitermes aburiensis (Sjostedt)	
Pericapritermes urgens (Silvestri)	
Ophiotermes grandilabius (Merson)	

Chapter 8
Utilitarian Use of Forest Insects

Introduction

The focus of much of this book is a compendium of important insects that negatively impact individual trees and forest ecosystems in tropical West Africa. There is however, quite another perspective on forest insects that we will address in this chapter. Insects contribute to the human enterprise in many and varied ways. We have long since recognized the ecological role of insects as pollinators, cyclers of nutrients, and natural enemies. This is quite an extensive discussion that needs to be addressed in a venue beyond the scope of this book. The utilitarian use of forest insects is however a subject we have decided to address in the second edition because there is great potential for insects to contribute to basic human needs and the economic development of tropical countries in West Africa.

Schabel (2006) has outlined several cases of insect use by humans in East Africa. Among these are insect collectors, human entomophagy (eating of insects by humans), silk production and beekeeping. To that list should probably be added the use of insects to produce lac used in various wood finishes and commercial dyes. For the final chapter we have elected to discuss insects as a potential resource to support ecotourism and the West African version of beekeeping to contrast with the East Africa situation as described by Schabel (2006). These topics will hopefully stimulate interest in our readers about ways to utilize forests and forest insects that are not directly related to the extractive use of forests.

Butterfly-Based Ecotourism

Throughout the world tourism is expanding as an enterprise because of increasing global wealth and a sincere interest on the part of many to enjoy and protect the natural environment. With growing global concern for sustainable development and conservation of resources there is increasing pressure to develop non-extractive strategies to provide societal benefit from the use of forests. Ecotourism is one example of a non-extractive approach to gaining human benefit while protecting nature.

Ecotourism Defined

Ceballos-Lascurain (1991) describes ecotourism as a mode of ecodevelopment that represents a practical and effective means of attaining social and economic improvement for all countries. The International Ecotourism Society defines ecotourism as responsible travel to natural areas that conserves the environment and sustains the well being of local people (Brandon 1996). Four types of travel are commonly considered as ecotourism: (1) nature-based tourism; (2) tourism that supports conservation efforts; (3) tourism that includes environmental awareness; and (4) tourism that is managed in a sustainable fashion (Brandon 1996). Ecotourism activities focus on experiencing and learning about nature and are designed to be low impact and locally focused to help indigenous people.

Ecotourism and Economic Development

Travel and tourism are major worldwide enterprises. These activities were estimated in 1995 to generate: (1) more than 10% of GDP; (2) 11.4% of global capital investment; and (3) contribute 655 billion USD to total tax payments (Brandon 1996). Tourism is viewed by many as a sustainable development approach because of the following features: (1) nonconsumptive use (watching birds, butterflies, and animals does not reduce their population); (2) ecotourism facilities are designed to be used in a sustainable fashion, i.e. low energy use, local labor and materials, use of renewable resources (wood vs. fossil fuel); (3) ecotourism activities are community based, revenue generated tends to stay in the local market, increases the economic diversification and stability of the local community. Certainly there are potential negative effects of tourism including the modification of local culture and the use of fossil fuel to complete travel to remote natural sites but the conventional view is that using the free market and tourism is the best way to focus private investment to achieve conservation objectives.

Why Butterflies?

There are a number of attributes of butterflies that make them highly suited for an ecotourism enterprise. Butterflies are generally viewed as highly aesthetic by the general public. To spend a few moments watching a butterfly suck nectar from a brilliant tropical flower while enjoying a walk in a tropical forest is a true delight. The fascination associated with the emergence of an adult butterfly from a pupa is an event that has captivated school children and adults for centuries. There is a global surge in butterfly zoos, exhibits, and gardens. This is exemplified by a number of books on butterfly gardening (Lewis 1995; Tekulsky 1985; Xerces Society 1990).

Butterflies can be artificially reared for release within protected enclosures. Rearing procedures for a few tropical butterflies are well established. For many species in Ghana the ability to rear them is limited by available life history information. Many butterfly-based ecotourism projects include butterfly ranching. This activity includes production of insect host material by local farms who participate in rearing butterflies that are then shipped to butterfly exhibits in developed countries. The value of rare or particularly attractive butterflies is sufficient to provide significant income to local farms and can represent a diversification of income that is important to subsistence farmers.

Another attribute of butterflies is that they can be attracted to live traps (Plate 63), bait stations (Plate 64), or planted gardens (Plate 65). Most tropical butterfly researchers use live traps baited with fermenting tropical fruits (banana is most common). These same baits can be placed in open containers that butterflies will locate and frequently utilize. Finally some nectar-producing flowers can be highly attractive to butterflies (Plate 66). By planting butterfly attractive plants in ornamental gardens it is possible to create open-air butterfly exhibits where visitors can observe butterflies that are free ranging.

There is significant conservation value associated with butterflies. There are esti-mated to be approximately 3,800 species of butterflies in Africa (D'Abrera 1980). There are about 730 forest-restricted butterfly species in Ghana (Larsen 2005). A list of butter-flies of Bobiri Forest Reserve (Ghana) is currently under preparation (Larsen, per com). Butterflies have been used to evaluate the role of sacred groves in conservation and the impact of forest fragmentation on biodiversity in Ghana (Bossart et al. 2006). Butterflies may well serve as good bioindicators of tropical forest biodiversity. The conservation of tropical biodiversity is a major global concern and what we learn about butterfly response may be key to the formation of conservation strategies (Bossart et al. 2006).

Bobiri Butterfly Sanctuary

The original concept for a butterfly ecotourism project in Ghana was formulated by Professor Wagner and Dr. Cobbinah in March, 1996 while they were jointly working on a research project funded by the International Tropical Timber Organization. In September, 1996, while on a US Fulbright Fellowship, Professor Wagner organized early butterfly collections (assisted by Dr. P.P. Bosu and Dr. E. Opuni-Frimpong), con-ducted butterfly attraction experiments, established butterfly gardens, and began improvements to the Bobiri Guest House (Plates 67 and 68). Bobiri was managed at that time by Jonas Osei Atiemo and Kingsley Safo. The Bobiri Guest House was a field research facility funded by the World Bank that was partially completed. Work on the guesthouse had been suspended due to the lack of funds. In February 1997, Professor Wagner gave a seminar to the staff of the Forestry Research Institute of Ghana outlining the concept of the butterfly sanctuary at Bobiri. In 1998, through the efforts of Dr. Cobbinah and Professor Wagner, the first US Peace Corps Volunteer, Gladstone "Mickey" Meyler was assigned to develop community support in the village of Kubease for the butterfly project. Two additional Peace Corps volunteers, Karen Moore and Jane Shalkey were instrumental in completing the Bobiri Guest House and launching the Butterfly Sanctuary over the next 6 years. From 1999–2004 the Bobiri Butterfly Sanctuary was a collaborative effort between the Nature Conservation Research Centre (Ghana), US Agency for International Development, Ghana Tourist Board, US Peace Corps, Forestry Research Institute of Ghana, and Northern Arizona University. Most of the basic butterfly taxonomy to support Bobiri was done by Torben Larsen. Ecological research on butterflies and ongoing butterfly identification were provided by Dr. Janice Bossart and her students. Educational posters were provided by Dr. Cobbinah, FORIG staff, Dr. Janice Bossart, and Northern Arizona University. Much of the early management of the butterfly sanctuary and guesthouse was done by Alfred Boakye. Contributing staff members included: Samuel Boateng, Evans Adrese, King Hiatorpe, Samuel Sarpong and Fatima Eshun. Alfred was the first technical officer to manage the Bobiri Butterfly Sanctuary and he brought many ecotourism principles to Bobiri.

The Bobiri Butterfly Sanctuary is the first of its kind in West Africa. The sanctuary is colocated with field research facilities within the Bobiri Forest Research about 4 km from the village of Kubease. The basic elements of the butterfly project are an overnight guesthouse, plantings of butterfly attracting plants, educational displays (Plate 69), gift shop, staff quarters, and of course tropical butterflies (Plate 71,72,73). At the site is also an arboretum, teak plantation, and a small demonstration of enrichment planting within a natural forest. About 450 species of butterflies have now been definitely recorded within Bobiri Forest (Larsen, unpublished). Study of the species composition and comparison with other forests in Ghana indicate that the forest contains about 600 of Ghana's 930 butterfly species. It is worth noting that there are many more butterflies in this small forest (54 km^2) than the whole of Western Europe and more than ten times in the countries of Northern Europe like Denmark and the Netherlands. The Bobiri butterfly sanctuary has been identified as a core centre for the proposed 2100 Ghana Butterfly Survey an international initiative which will show whether significant changes have taken place during the century. Butterflies records since the project inception in 1996 will provide the baseline data for 2100 surveys. Butterfly collections for research and voucher purposes are maintained at the Forestry Research Institute of Ghana.

The development of a significant butterfly-based ecotourism enterprise in Ghana is yet to be realized. Bobiri Forest and Butterfly Sanctuary have been selected as one of five global canopy study site. The Bobiri Butterfly Sanctuary still needs substantial development before it will become a source for economic development for the nearby village of Kubease. New sources of capital investment need to be identified. Modern principles of hospitality management must be employed. Finally, there is extensive need for research on the basic butterfly ecology and rearing of butterflies. The full potential for butterfly based ecotourism can only be achieved when all of these are moved forward in unison.

Apiculture

Harvesting of honey from wild bee colonies has been practiced in West Africa for many years. Under the traditional system of collecting honey from wild colonies, fire was set to the tree containing the honeycomb to drive out the bees before harvesting was done. By this method, however, many bees were killed and the colony usually destroyed. This approach is sometimes described as "bee-killing," as opposed to "bee-keeping," where bees are kept in specially designed hives and actively managed to obtain optimum yield of honey and other bee products. In between the two extremes, there is "bee-having," which usually involves housing bees in simple enclosures or receptacles such as tree trunks, gourds, and straw or clay pots, with rarely any management intervention except perhaps occasional inspection to determine how much honey has been produced.

Apiculture or contemporary beekeeping in Ghana began with a failed attempt by the government to raise imported European bees in the 1960s. Not much was done after this initial setback until in the late 1970s and early 1980s when a few enthusiastic individuals began to maintain a few hives, working with the tropical honey bee.

Beekeeping with African Bees in the Tropics

The tropical honeybee *Apis mellifera adansonii* (Scute-Lata) is the preferred bee strain for beekeepers in Africa. It has a slightly smaller body than its European counterpart.

Although they are generally more aggressive and more difficult to manage they are better adapted to the tropical environment than European bees. Some have suggested this aggressive behavior to have developed as a result of harsh environmental conditions coupled with years of hostile encounters with humans (frequent burn out of bees from trees to harvest honey). Tropical bees are particularly aggressive during the hot hours of the day, and readily abscond when disturbed, or migrate when environmental conditions change substantially from optimum conditions. When disturbed, tropical bees can follow their victim for over 400 m, while European bees may follow for just about 50 m (Adjare 1990). By this nature, tropical bees require a copious amount of smoke given at frequent intervals to calm them down when hives are opened. The alarm pheromone of tropical honey bees is more powerful that those of European bees.

Beehives

In 1851 Reverend Lorenzo Langstroth of Philadelphia, USA, discovered the principle of "bee space" which transformed beekeeping the world over. Bee space is simply the amount of space that bees need to pass easily between two surfaces. He found out that bees require a crawl space of between 6–9 mm. If the space was less than required the bees would seal it with propolis. On the other hand, if there was more space, then the bees would construct a comb in that space. The discovery of the principle of bee space was the bases for the invention of the modern day movable frame hive, which is often called the Langstroth hive (Dadant 1976).

Following Langstroth's discovery many other frame hives were invented based on the principle of the bee space (Morse 1994). The invention of top-bar hives, in particular, changed the nature of beekeeping in developing countries. This is a modified version of the Langstroth hive in which bars replace the movable frames. Bees build combs that hang vertically from the center of the top bars. There are several types of top bars including the V-shaped top-bar, groove top-bar, and the pointed starter. However, in all these a critical top bar width of 32 mm should be observed. The width of the comb built by the tropical honey bee and attached to the center of the top bar is 25 mm, which leaves a space of 3.5 mm on either side of the 32 mm top bar. Thus, a bee space of 7 mm is created between adjacent top bars, perfect for bee movement and permitting removal and hive inspection.

Many types of top-bar hives are in use in developing countries, however, in Ghana, the Kenyan top bar hive (KTBH) is the most frequently used. It is made by local carpenters and craftsmen using available equipment. Though frequently made of wood, it can also be made from straw, woven reeds, bamboo with mud, or metal containers depending on the climate and the available local materials. In addition to the top bar hive, bee suits, veils, smokers, and other basic equipment can be obtained from local sources.

Stingless Bees (Other Bees; Sweat Bees)

Getting stung by bees is, unfortunately, as much a part of beekeeping as is eating honey. The fear of getting stung repeatedly by bees is a major disincentive to prospective beekeepers. For individuals who have allergic reactions to bee stings this may be a good enough justification to avoid beekeeping. However, there are other species of honey-producing bees that do not sting intruders. Domesticating stingless bees for honey production is vastly becoming popular. Although they produce less honey than true honey bees, keeping stingless bees can help stimulate the interest needed for beginners. In tropical Africa species of *Trigona* and *Melipona* are known examples of honey-producing stingless bees. Though they do not sting, they may defend their colonies by biting and secreting irritating substances to keep off intruders.

Production and Marketing of Honey in Ghana

An estimated 60% of honey on the Ghanaian market is harvested from wild bee colonies. According to Aidoo (2005) there are about 5,000 beekeepers in Ghana with an average of five hives per person, with a total honey production of about 7 tons annually. Honey harvested from the wild is usually black and sooty, and of very poor quality because of the harvesting and processing methods used. In Ghana beekeepers usually use cold extraction methods, honey press, or the solar wax extractor to extract honey from the comb. Improper use of the latter can reduce the quality of the honey as a result of overheating (Aidoo 2005).

The period of honey production in Ghana is from April through November. During the period, bulk buyers from urban centers visit production centers to purchase honey from individual beekeepers or groups. They in turn sell to bulk users or retailers in the cities. Honey is usually sold in the open markets or in supermarkets. At least two established beekeeping centers, The Honey Center at Saltpond and Farm Bee in Accra produce and package raw honey and other bee products for sale. A few supermarkets also import honey from Europe for sale mainly to foreign consumers. Average price of locally produced honey is around US $6/kg. Imported honey available only in a few selected supermarkets averages about US $20/kg (Aidoo 2005).

Honey produced in Ghana is used in traditional medicinal preparations, as food sweeteners, and to a lesser extent as ingredients in locally made cosmetics. Sources of information on beekeeping and technical assistance in Ghana include the Forestry Research Institute of Ghana (FORIG), the Technology Consultancy Center (TCC) of the Kwame Nkrumah University of Science and Technology (KNUST), and the Sasakawa Center, University of Cape Coast (UCC). Many other enthusiastic beekeepers and beekeeping associations sometimes supported by such development organizations as the USAID (The United States Agency for International Development), ADRA (Adventist Relief Agency), World Vision International, and TECHNOSERVE, can be found throughout the country.

References

Adjare, S.O. 1990. Beekeeping in Africa. Food and Agriculture Organization Agricultural Services Bulletin 68/6.

Agounke, D., U. Agricola, and H.A. Bokonon-Ganta. 1988. *Rastrococcus invadens* Williams (Hemiptera: Pseudococcidae), a serious exotic pest of fruit trees and other plants in West Africa. Bulletin of Entomological Research 78: 695–702.

Aidoo, K. 2005. The honey trade in Ghana. Proceedings of Bees for Development Honey Trade Workshop held in Dublin, Ireland in August 2005.

Akanbi, M.O. 1986. Observations of the biotic factors affecting the populations of *Epicerura pulverulenta* Hampson (Lepidoptera: Notodontidae). Insect Science and its Application 7(6): 785–789.

Akanbi, M.O. 1990. Biology, behaviour and seasonal fluctuations of *Epicerura pulverulenta* Hampson (Lepidoptera: Notodontidae). Discovery and Innovation 2(2): 85–90.

Akanbi, M.O., I.B. Alebiosu, and A.A. Alabi.1990. An outbreak and control of *Aonidiella* species (Hemiptera: Diaspididae) on *Azadiratcha indica* (neem) in Nigeria. In: Hutacharan, C, K.G. MacDicken, M.H. Ivory, and K.S.S. Nair (eds), Proceedings of the IUFRO Workshop on pests and diseases of Forest Plantations, RAPA, FAO, Bangkok.

Alder, D. 1989. Natural forest increment, growth and yield. In: Ghana Forestry Inventory Proceedings. ODA/Ghana Forestry Department, Accra, pp 47–52.

Ande, A.T. and J.O. Fosaranti. 1997. Life history notes for the pallid emperor moth, *Cirina forda* Westwood (Lepidoptera: Saturniidae). Journal of Lepidopterists Society 51: 269–271.

Anonymous 1957. Report of the West Africa Timber Borer Research Unit for 1953–1955. Eyre and Spottiswoode, Chiswick press, London, 44 pp.

Anonymous 1959. Report of the West African Timber Borer Research Unit for 1955–1958. Eyre and Spottiswoode, Chiswick Press, London, 62 pp.

Anonymous 1960. Report of the West African Timber Borer Research Unit 1958–1959. Eyre and Spottiswoode, Chiswick Press, London, 45 pp.

Anonymous 1961. Fourth Report of the West African Timber Borer Research Unit. Eyre and Spottiswoode, Chiswick Press, London, 96 pp.

Anonymous 1962. Fifth Report of the West African Timber Borer Research Unit. Eyre and Spottiswoode, Chiswick Press, London, 96 pp.

Anonymous 1970. The dry-wood termite (*Cryptotermes havilandi*). Information Sheet, Building and Road Research Institute, Kumasi-Ghana No. 8 (Second Series), 5 pp.

Ampong, F.F.K. 1977.Preliminary investigations into the wood boring activities of the nymphal stages of *Povilla adjusta* Navas at Nzulezo, Western Ghana. Forest Products Research Institute Technical Note No. 24, 12 pp.

Apetorgbor, M., F. Mancini, E. Turco, J.R. Cobbinah, and A. Ragazzi. 2001. The involvement of fungal pathogens in dieback-decline of *Milicia excelsa* saplings in plantations. Zeitschrift-fur-Pflanzenkrankheiten-und-Pflanzenschutz 108(6): 568–577.

Ashiru, M.O. and B. Momodu. 1981. Wood-boring habits of *Eulophonotus obesus* on *Triplochiton scleroxylon*. Malaysian Forester 44(4): 473–481.

Ashiru, M.O. and B. Momodu. 1986. Control of *Eulophonotus obesus* (Lepidoptera: Cossidae): A borer of *Triplochiton scleroxylon* K. Schum. Malaysian Forester 49(2): 198–204.

Ashiru, M.O. 1988a. The food value of the larvae of *Anaphe venata* Butler (Lepidoptera: Notodontidae). Ecology of Food and Nutrition 22–313–320. Forestry Research Institute of Nigeria, Ibadan, Nigeria.

Ashiru, M.O. 1988b. The frequency distribution of eggs and larvae of *Anaphe venata* Butler (Lepidoptera: Notodontidae) on *Triplochiton scleroxylon* K. Schum. Insect Science Applications 9(5): 587–592.

Atuahene, S.K.N. 1970. The economic effect of insect pests on the timber industry in Ghana, Forest Products Research Institute Technical Newsletter 4(4): 4–8.

Atuahene, S.K.N. 1972. Preliminary studies on woodwool resistance to termite attack in Ghana. Forest Products Research Institute Technical Newsletter 6(3 and 4): 1–10.

Atuahene, S.K.N. 1975. The longhorn borer, *Analeptes trifasciata Eucalyptus* spp. in the southern Savanna woodlands of Ghana. 9th Biennial Conference of the Ghana Science Association, Legon.

Atuahene, S.K.N. 1976. Incidence of *Apate* spp. (Coleoptera: Bostrychidae) on young forest plantations species in Ghana. Ghana Forestry Journal 23: 29–35.

Atuahene, S.K.N. 1977. The occurrence and pests status of *Analeptes trifasciata* F. (Coleoptera: Lamiinae) on *Eucalyptus* species in the coastal thicket of South Central Ghana. Ghana Forestry Journal 3: 53–59.

Atuahene, S.K.N. 1983. The biology of *Lamprosema lateritialis* Hampson (Lepidoptera: Pyralidae), a pest of Afrormosia (*Pericopsis elata*) (Harms) Van Meeuwen in Ghana. Ph.D. Thesis, University of Ghana.

Atuahene, S.K.N. 1989. Effect of defoliation on *Pericopsis elata* (Harms) van Meeuwen by the leaf-trying moth *Lamprosema lateritialis* Hamps. Forest Products Research Institute Technical Bulletin 8.

Atuahene, S.K.N. and S.A. Nkrumah. 1977. Sampling populations of ambrosia beetles in the high forest belt by means of the Tilley Storm lamp. Forest Products Research Institute Technical Bulletin 1(1): 37–49.

Atuahene, S.K.N. and H. Doppelreiter. 1982. Histopathological observations on *Beauveria bassiana* in larvae of *Lamprosema lateritialis* (Lepidoptera: Pyralidae). Zeitshrift fur Angewandte Entomologie 93: 456–463.

Atuahene, S.K.N. and D. Souto. 1983. The rearing and biology of the mahogany shoot borer *Hypsipyla robusta* Moore (Lepidoptera: Pyralidae) on an artificial medium. Insect Science Application 4(4): 319–325.

Barbosa, P. and M.R. Wagner. 1989. Introduction to Forest and Shade Tree Insects. Academic Press, San Diego, 639 pp.

Berryman, A.A. 1986. Forest Insects: Principles and Practice of Population Management. Plenum Press, New York, 279 pp.

Bokonon-Ganta, A.H., H. de Groote, and P. Neuenschwander. 2002. Socio-economic impact of biological control of mango mealybugy in Benin. Agriculture, Ecosystems and Environment 93: 367–378.

Borror, D.J. and D.N. Delong. 1964. An Introduction to the Study of Insects. Holt, Rinehart and Winston, New York.

Bossart, J.L., E. Opuni-Frimpong, S. Kuudaar, and E. Nkrumah. 2006. Richness, abundance and complementarity of fruit-feeding butterfly species in relict sacred forests and forest reserves in Ghana. Biodiversity and Conservation 15: 333–359.

Bosu, P.P. 1999. Survey and evaluation of natural enemies of the iroko gall bug, *Phytolyma lata* (Homoptera: Psyllidae) in Ghana. M. Phil. Thesis, University of Science and Technology, Kumasi, Ghana, p. 89.

Bosu, P.P.,S. Adu-Bredu, and E.E. Nkrumah. 2004. Observations of insect pest activities within selected nurseries in Ashanti, Ghana. 154–162. Pp 48–69p. In: Pest Management in Tropical Plantations (eds) Cobbinah, J.R., D.A. Ofori, and P.P. Bosu. Proceedings of an International Workshop held in Kumasi, Ghana, 21–23 July 2004, 185 pp.

Bosu, P.P., J.R. Cobbinah, and M.R. Wagner. 2000. Feasibility of biological control of *Phytolyma lata* in Africa pp 77–90. In: Research Advances in Restoration of Iroko as a Commercial

Species in West Africa (eds) Cobbinah, J.R. and M.R. Wagner. Proceedings of an International Workshop held in Kumasi, Ghana, 15–16 November 2000, 134 pp.

Bosu, P.P., J.R. Cobbinah, E. Frempong, J.D. Nichols, and M.R. Wagner. 2004. Evaluation of indigenous parasitoids of the iroko (*Milicia excelsa*) gall bug, *Phytolyma lata* Scott (Homoptera: Psyllidae). Ghana Journal of Forestry 15 and 16 1–12.

Bosu, P.P., J.R. Cobbinah, E.E. Nkrumah, E. Frempong, and M.R. Wagner. 2003. A quantitative evaluation of indigenous predators on the iroko gall bug *Phytolyma lata* (Scott) (Homoptera: Psyllidae) in Ghana. Ghana Journal of Forestry 11(2): 39–50.

Bosu, P.P., J.R. Cobbinah, J.D. Nichols, E.E. Nkrumah, and M.R. Wagner. 2006. Survival and growth of mixed plantations of *Milicia excelsa* and *Terminalia superba* 9 years after planting in Ghana. Forest Ecology and Management 233: 352–357.

Bradley, J.D. 1968. Descriptions of two new genera and species of Phyticinae associated with *Hypsipyla robusta* (Moore) on Meliaceae in Nigeria (Lep: Pyralidae). Bulletin Entomological Research 57: 605–613.

Brandon,K.1996.Ecotourism and Conservation: A Review of Key Issues. World Bank Environment Department Paper. No. 35, 69 pp.

Browne, F.G. 1961a. Preliminary observations on *Doliopygus dubius* (Samps.) (Coleopotera: Platypodidae) pp. 25–30. In: Fourth Report of the West African timber Borer Research Unit. Eyre and Spottiswoode, Chiswick Press, London.

Browne, F.G. 1961b. The genetic characteristics, habits and taxonomic status of *Premnobius* Eichh. (Coleoptera: Scolytidae) pp. 45–51. In: Fourth Report of the West African Timber Borer Research Unit. Eyre and Spottiswoode, Chiswick Press, London.

Browne, F.G. 1962a. The emergence, flight and mating behaviour of *Doliopygus conradti* (strohm.) (Coleoptera: Platypodidae) pp. 21–27. In: Fifth Report of the West African Timber Borer Research Unit. Eyre and Spottiswoode, Chiswick Press, London.

Browne, F.G. 1962b. Notes on *Xyleborus ferrugineus* F. (Coleoptera: Scolytidae) pp. 47–55 In: Fifth Report of the West African Timber Borer Research Unit. Eyre and Spottiswoode, Chiswick Press, London.

Browne, F.G. 1963. Notes on the habits and distribution of some Ghanaian bark beetles and ambrosia beetles (Coleoptera: Scolydtidae and Platypodidae). Bulletin of Entomological Research 54: 229–266.

Browne, F.G. 1964. Types of ambrosia beetle attack on living trees in tropical forests. Proceeding XIIth International Congress Entomology, London, 680 pp.

Browne, F.G. 1968. Pests and Diseases of Forest Plantation Trees: An Annonated List of the Principal Species Occurring in the British Commonwealth. Clarendon Press, Oxford, 1330 pp.

Brunck, F. and B. Mallet. 1993. Problems relating to pests of mahogany in Cote d'Ivoire. Bois et Forets des Tropiques 237: 9–29.

Ceballos–Lascurain, H. 1991. Tourism, ecotourism and protected Areas. Parks 2(3): 31–35.

Cobbinah, J.R. 1972a. A survey of nursery pests. Planning Branch Entomology Report 1 Ghana Forestry Department.

Cobbinah, J.R. 1972b. *Hypothenemus pussilus*, a shoot borer of *Tectona grandis, Terminalia ivorensis, Cedrela odorata* and *Gmelina arborea*. Planning Branch Entomology Report 2. Ghana Forestry Department.

Cobbinah, J.R. 1973. Natural variation in susceptibility of four species of *Eucalyptus* to *Strepsicrates rotha – Meyr* (Lepidoptera: Tortricidae). Planning Branch Entomology Report 4, Ghana Forestry Department.

Cobbinah, J.R. 1983. Evaluation of five insecticides for the control of *Phytolyma lata* (Homopotera: Psyllidae) populations. Forest Products Research Institute Technical Bulletin 3: 35–41.

Cobbinah, J.R. 1986. Factors affecting the distribution and abundance of *Phytolyma lata* (Homoptera: Psyllidae). Insect Science Application 7(1): 111–115.

Cobbinah, J.R. 1988. The biology, seasonal activity and control of *Phytolyma lata*. IUFRO Regional Workshop on Pests and Diseases of Forest Plantations, June 5–10. Bangkok, Thailand.

Cobbinah, J.R. 1993. Snail Farming in Ghana. Forestry Research Institute of Ghana, 46 pp.

Cobbinah, J.R. and M.R. Wagner. 1995. Phenotypic variation in *Milicia excelsa* to attack by *Phytolyma lata* (Psyllidae). Forest Ecology and Management 75: 147–153.

Cobbinah, J.R. and M.R. Wagner (eds). 2000. Research Advances in Restoration of Iroko as a Commercial Species in West Africa. Proceedings of an International Workshop held in Kumasi, Ghana, 15–16 November 2000, 134 pp.

Cornelius, J.P. 2001. The effectiveness of pruning in mitigating *Hypsipyla grandella* attack on young mahogany (*Swietenia macrophylla* King) trees. Forest Ecology and Management 148: 287–28.

Coulson, R.N. and J.A. Witter. 1984. Forest Entomology: Ecology and Management, Wiley, New York, 669 pp.

D'Abrera, B. 1980. Butterflies of the Afrotropical Region. Lansdowne Editions, Melbourne, Australia.

Dadant, C.P. 1976. First Lessons in Beekeeping (Revised Edition). Dadant & Sons, Hamilton, Illinois. 130p.

Dajoz, R. 2000. Insects and Forests: The Role of Diversity of Insects in the Forest Environment. Intercept Publishing, London, 668 pp.

Danso, L.K. 1970. Seed production by *Triplochiton sxleroxylon* K. Schum (Wawa). Forest Products Research Institute Technical Newsletter 4(1): 21–22.

Duffy, E.A.J. 1957. A monograph of the immature stages of African timber beetles (Cerambycidae). Printed by order of the Trustee of the British Museum, London.

Dwomoh, E.A. 2003. Insect species associated with sheanut tree (*Vitellaria paradoxa*) in Northern Ghana. Tropical Science 43: 70–73.

Dwomoh, E.A., Akrofi, A.Y., and S.K. Ahadzi. 2004. Natural enemies of the sheanut defoliator, *Cirina forda*. Tropical Science 44: 124–127.

Eidt, D.C. 1963. A survey of insect pests of indigenous trees in plantations and nurseries. Food and Agriculture Organization. United Nations Report No. 1775. Rome, 68 pp.

Eidt, D.C. 1965a. The Opepe shoot borer, *Orygomophora mediofoveata* Hmps. (Lepidoptera: Noctuidae), a pest of *Nauclea diderrichii* in West Africa Commonwealth. Forestry Review 44(2) No. 120: 123–125.

Eidt, D.C. 1965b. Description of the larva of *Orygmophora mediofoveata* Hampson (Lepidoptera: Noctuidae). Reprinted from Canadian Entomology 612–617.

Entwistle, P.F. 1963. A note on *Eulopohonotus mymeleon* Fld. (Lepidoptera: Cossidae) a stem borer of cocoa in West Africa. Bulletin Entomological Research 54: 1–3.

Entwistle, P.F. 1964. In-breeding and arrhenotoky in the ambrosia beetle *Xyleborus compactus* Eich. Proceedings Royal Entomological Society London 39: 83–88.

Farid, A. 1994. Study on bio-ecology and control of *Aonidiella orientalis* in Jiroft and Hormozgan. Applied Entomology and Phytopathology 61: 29, 96–105.

Flyod, R.B. 2001. International Workshop on Hypsipyla shoot borers in Meliaceae: General Conclusions and research priorities. In. Hypsipyla shoot borers in Meliaceae. Proceedings of an International Workshop held at Kandy, Sri Lanka, 20–23 August 1996. pp. 183–187.

Foahom, B. 1994. Oviposition behaviour in *Godasa sidae*, a defoliating caterpillar of *Mansonia altissima*. Cahiers Agricultures 3: 315–317.

Foahom, P. and Du Merle, P. 1993. First data on the biology of *Godasa sidae* (Fab.) (Lep., Noctuidae), a pest of *Mansonia altissima* (Sterculiaceae) in Cameroun. Journal of Applied Entomology 116: 284–293.

Food and Agriculture Organization. 2005. State of the World's Forests. FAO, United Nations, Rome, 153 pp.

Forsyth, J. 1966. Agricultural Insects of Ghana. Ghana University Press, Accra, 163 pp.

Ghauri, M. 1962. The morphology and taxonomy of male scale insects (Homoptera, Coccoidea). British Museum (Natural History). Adlard & Son, Dorking, UK, 221 pp.

Glover, P.M. 1933. *Aspidiotus (Furcaspis) orientalis* Newstead, its economic importance in lac cultivation and its control. Indian Lac Research Institute Bulletin 16: 1–22. Handbook 148. Brooklyn, New York, 111 pp.

Gyimah, A. 1984. Storage of *Mansonia altissima* seeds. Forest Products Research Institute Technical Bulletin 4: 16–20.

Hall, J.B. and M.D. Swaine. 1976. Classification and ecology of closed canopy forest in Ghana. Journal of Ecology 64: 913–951.

Hall, J.B. and M.D. Swaine. 1981. Distribution and Ecology of Vascular Plants in a Tropical Rain Forest: Forest Vegetation of Ghana. W. Junk Publishers, The Hague, 383 pp.

Harris, W.V. 1955. Termites and forestry. Empire Forestry Review 34: 156–166.

Harris, W.V. 1961. Termites – Their Recognition and Control. Longmans, Green, London.

Harris, W.V. 1964. Termites – Their Recognition and Control. Longmans, Green, London. 187.

Hassani, A.E. and J. Messaoudi. 1986. Les ravageurs des cones et graines de conifers et leur distribution au Maroc pp. 5–14. In: A. Roques (ed.) Proceedings of the Second Conference of the Cone and Seed Insects Working Party – IUFRO, 312 pp.

Hauxwell, C., C. Vargas, and E. Opuni-Frimpong. 2001. Entomopathogens for control of *Hypsipyla* spp. In: Hypsipyla shoot borers in Meliaceae. Proceedings of an International Workshop held at Kandy, Sri Lanka, 20–23 August 1996. pp. 131–139.

Jones, N. 1969. A description of fruit of *Terminalia ivorensis*. Forest Products Research Institute Technical Newsletter 3(3*4): 11–15.

Jones, N. and H.B. Damptey. 1969. Kumasi seed store. Forest Products Research Institute Technical Note 8: 6.

Jones, N. and J. Kudler. 1968. A report of the preliminary work on the influence of weevil attack on the germination of *Terminalia ivorensis*. Forest Products Research Institute Technical Newsletter 2(2): 11–14.

Jones, N. and J. Kudler. 1969. Some particulars concerning weevil attack on the fruits of *Terminalia ivorensis*. Forest Products Research Institute Technical Newsletter 3(3&4): 7–10.

Jones, T. 1959a. The major insect pests of timber and lumber in West Africa. West African Timber Borer Research Unit Technical Bulletin No. 1. Eye and Spottiswoode Ltd., Chiswick Press, London, 20 pp.

Jones, T. 1959b. Ambrosia beetles (Scolytidae) their biology and control in West Africa. West African Timber Borer Research Unit Technical Bulletin No. 2. Eyre and Spottiswoode, Chiswick Press, London, 14 pp.

Kanga, L. and G. Fediere. 1991. Towards integrated control of *Epicerura pergrisea* (Lepidoptera: Notodontidae), defoliator of *Terminalia ivorensis* and *T. superba*, in the Côte d' Ivoire. Forest Ecology and Management, 39(1–4): 73–79.

Khalaf, J. and M. Sokhansanj. 1994. Bioecological studies on orientalis yellow scale (*Aonidiella orientalis* New.) and its control by integrated methods in Fars province. Applied Entomology and Phytopathology 60: 53–59 (Persian), 11–12 (English).

Kudler, J. 1967a. Problems of forest insect defoliators in Ghana. Proceedings XIV International Union of Forestry Research Organizations Congress, Munich. Papers V Section 24: 618–627.

Kudler, J. 1967b. *Lamprosema lateritalis* Hamps, a new serious defoliator of Kokrodua, *Afrormosia elata* Harms. Forest Products Research Institute Newsletter 4: 3–5.

Kudler, J. 1968. *Diclidophlebia eastopi* (Vondracek), a psyllid harmful to *Triplochiton scleroxylon* (K. Schum). Forest Products Research Institute Technical Newsletter 2(3*4): 11–14.

Kudler, J. 1970a. Insect attacks on *Sesbania grandiflora* (L.) Poir. Forest Products Research Institute Technical Newsletter 4(4): 9–12.

Kudler, J. 1970b. Termite attack and protection of *Euclayptus* plantations in Ghana. Ghana Journal of Agricultural Science 3: 39–44.

Kudler, J. 1970c. Attack of *Menechamus* sp. (Col., Curculionidae) on mature fruits of *Guarea cedrata*. Forest Products Research Institute Technical Newsletteer 4(3): 7–10.

Kudler, J. 1971. Observations on *Tridesmodes ramiculata* Warr (Lepidoptera: Thyrididae) a shoot borer of *Terminalia ivorensis*. A Chev. Forest Products Research Institute Newsletter 5(1): 6–10.

Kudler, J. 1978. An outline of forest and wood product entomology in Ghana. Silvaecultura Tropica Et Subtropica 6: 15–43.

Kudler, J. and N. Jones. 1970. Pests reducing the quality of *Triplochiton scleroxylon* K. Schum fruits in flowering years. Forest Products Research Institute Newsletter 4(1): 16–20.

Kudler, J. and S.O.A. Quaynor. 1971. A note on *Erythrina addisoniae* Hutch & Dalz and its shoot borer, *Terastria reticulate* Gn. (Lep. Pyralidae), Forest Products Research Institute Technical Newsletter 5(3): 1–4.

Lale, N.E.S. 1998. Neem in the conventional Lake Chad Basin area and the threat of oriental yellow scale insect (*Aonidiella orientalis* Newstead) (Homoptera: Diaspididae). Journal of Arid Environments 40: 191–197.

Larsen, T.B. 2005. The Butterflies of West Africa. Apollo Books, Svendborg, Denmark.

Lawson, G.W. 1986. Plant Life in West Africa. Ghana University Press, Accra, 138 pp.

Lepesme, P. 1953. Coleopteres Cerambycidae (Longicornes) de Cote d'Ivoire-Catalogues, Institut francais D'Afrique noire. Vol. XI, Ifran-Darkar.

Lewis, A. (ed.). 1995. Butterfly Gardens. Brooklyn Botanical Garden Publication, Handbook 148. Brooklyn, New York, 111 pp.

Mahroof, R.M., C. Hauxwell, J.P. Edirisinghe, A.D. Watt, and A.C. Newton. 2002. Effects of artificial shade on attack by the mahogany shoot borer. Agricultural and Forest Entomology. 4(4): 283–292.

Mantanmi, B.A. 1988. An outbreak of suspected virosis amongst natural populations of *Epicerura pulverulenta* Hampson (Lepidoptera: Notodontidae).

Mensa-Bonsu, A. 1973. The principles of pest control as applied to the control of termites. Forest Products Research Institute Technical Newsletter 7(3 and 4): 1–7.

Morse, R.A. 1994. The Complete Guide to Beekeeping. The Countryman Press, Woodstock, Vermont. 207p.

Newton, A.C., J.P. Cornelius, J.F. Mesen, and R.R.B. Leakey. 1995. Genetic variation in apical dominance of *Cedrela odorata* seedlings in response to decapitation. Silvae Genetica 44: 146–150.

Newton, A.C., J.P. Cornelius, J.F. Mesen, E.A. Corea, and A.D. Watt. 1998. Variation in attack by the mahogany shoot borer, *Hypsipyla grandella* (Zeller) in relation to host growth and phenology. Bulletin of Entomological Research, 88: 319–326.

Newton, A.C., J.P. Cornelius, J.F. Mesen, E.A. Corea, and A.D. Watt. 1999. Genetic variation in host susceptibility to attack by the mahogany shoot borer, *Hypsipyla grandella* (Zeller). Agricultural and Forest Entomology 1: 11–18.

Newton, A.C., J.P. Cornelius, P. Baker, A.C.M. Gillies, M. Hernandez, S. Ramnarine, J.F. Mesen, and A.D. Watt. 1996. Mahogany as a genetic resource. Biological Journal of the Linnean Society 122: 61–73.

Nichols, J.D., D.A. Ofori, M.R. Wagner, P. Bosu, and J.R. Cobbinah.1999. Survival, growth and gall formation by *Phytolyma lata* on *Milicia excelsa* established in mixed-species plantations in Ghana. Agricultural and Forest Entomology 1 137–141.

Nkansa-Kyere, M. 1972. Exotic forest trees of Ghana. Forest Products Research Institute Information Bulletin 5: 2 pp.

Ocloo, J.K. 1983. A comparative study of the protection offered to wood samples by permethrin, dieldrin and lindane against damage by subterranean termites to fungi. International Journal of Wood Preservation 3(1): 31–33.

Ocloo, J.K. and M.B. Usher. 1980. The resistance of 85 Ghanaian hardwood timbers to damage by subeterranean termites. Special Technical Publications 691.

Odebiyi, J.A., A.A. Omoloye, S.O. Bada, R.O. Awodoyin, and P.I. Oni. 2003. Spatial Distribution, pupation behaviour, and natural enemies of *Cirina forda* Westwood (Lepidoptera: Saturniidae) around its host, the sheanut tree. Insect Science and its Application 23(3): 267–272.

Ofori, D.A. 2001. Genetic diversity and its implications for the management of *Milicia* species. Ph.D. Thesis. University of Aberdeen. p. 158.

Ofori, D.A. and J.R. Cobbinah. 2007. Integrated approach for conservation and management of genetic resources of *Milicia* species in West Africa. Forest Ecology and Management 238: 1–6.

Ofori, D.A., J.R. Cobbinah, and E. Opuni-Frimpong. 2004. Development of an integrated strategy for reduction of shoot borer impact on mahogany in Ghana, pp 48–69 p. In: Pest Management in Tropical Plantations (eds) Cobbinah, J.R., D.A. Ofori, and P.P. Bosu. Proceedings of an International Workshop held in Kumasi, Ghana, 21–23 July 2004, 185 pp.

Ofori, D.A., E., Opuni-Frimpong, and J.R. Cobbinah. 2007. Provenance variation in *Khaya* spp for growth and resistance to the shoot borer *Hypsipyla robusta* Forest Ecology and Management 242: 438–443.

Ofori, D.A., J.R. Cobbinah, M.D. Swaine, A.H. Price, C. Leifert, and S. Adu-Bredu. 2000. Screening for genetic resistance to *Phytolyma lata* in *Milicia* species (odum/Iroko) using DNA fingerprinting and phenotypic characters pp 47–53. In: Research Advances in Restoration of Iroko as a Commercial Species in West Africa (eds) Cobbinah, J.R. and M.R. Wagner. Proceedings of an International Workshop held in Kumasi, Ghana, 15–16 November 2000, 134 pp.

Okoro, O.O. and A.O. Dada. 1987. Forest seed problems of Nigeria. pp. 225–237. In: Proceedings of the International Symposium on Forest Seed Problems in Africa (eds) Kampra, S.K. and R.D. Ayling. International Union of Forestry Research Organizations Report 7, 399 pp.

Opuni-Frimpong, E, D.F. Karnosky, A. Storer, and J.R. Cobbinah. 2004. Conserving African mahogany diversity: provenance selection and silvicultural systems to develop shoot borer tolerant trees pp 71–84. In: Pest Management in Tropical Plantations (eds) Cobbinah, J.R., D.A. Ofori, and P.P. Bosu. Proceedings of an International Workshop held in Kumasi, Ghana, 21–23 July 2004, 185 pp.

Opuni-Frimpong, E., D.F. Karnosky, A. Storer, J.R. Cobbinah, and D.A. Ofori. 2005. Development of an integrated management strategy to reduce the impact of *Hypsipyla* species (Lepidoptera: Pyralidae) on African mahogany. International Forestry Review v.7(5) p. 86. Parks 2(3): 31–35.

Orr, R.L. and A. Osei-Nkrumah. 1978. Progress Report on *Phytolyma lata* Hemiptera: Psyllidae. Forest Products Research Institute, Kumasi, Ghana.

Osafo, E.D. 1970. The development of silvicultural techniques applied to natural forests in Ghana. FPRI Technical No. 13. Kumasi, Ghana.

Osisanya, E.O. 1968. The taxonomy and biology of two *Diclidophlebia* species (Homoptera: Psyllidae) on *Triplochiton scleroxylon* in Nigeria. Ph.D. Thesis, University of Ibadan, Nigeria.

Ouedraogo, A.S. and J.A. Verwey, 1987. Forest tree seed problems in Burkina Faso (Sahelian and Soudanian Regions) pp. 238–249. In: Proceedings of the International Symposium on Forest Seed Problems in Africa (eds) Kamra, S.K. and R.D. Ayling. International Union of Forestry Research Organizations Report 7, 399 pp.

Quao, H.N.O. 1965. The control of termites in building. Ghana Journal of Science 5(1): 74–77.

Parren, M.P.E. and de Graaf, N.R. 1995. The quest for natural forest management in Ghana, Cote d'Ivoire and Liberia. Tropenbos Series 13, Wageningen, The Netherlands, 199 pp.

Parry, M.S.1956. Tree planting practices in tropical Africa. FAO Forestry Development Paper 8, 298 pp.

Rasplus, J.Y. 1988. La communaute parasitaire des Coleopteres seminivores de Legumineuses dans une mosaique foret-squane en Afrique del Ouest (Lamto-Cote d'Ivoire). These Doctorat. Devant l'Universite D'Orsay-Paris XI.

Ritchie, J.M. 1987. Insect biosystematic services in Africa: current status and future prospects. Insect Science Applications 8(4/5/6): 425–432.

Roberts, H. 1961a. Preliminary survey of the activity of timber pests in Takoradi Harbor, Ghana pp. 39–43. In: Fourth Report of the West African Timber Borer Research Unit. Eyre and Spottiswoode, Chiswick, London.

Roberts, H. 1961b. *Analeptes trifasciata* F. a longhorn borer that attacks members of the Bombacaceae in Ghana pp. 59–66. In: Fourth Report of the West African Timber Borer Research Unit. Eyre and Spottiswoode, Chiswick, London.

Roberts, H. 1962a. A description of the developmental stages of *Trachyostus aterrimus* (Schauf.), A West African Platpodid, and some remarks on its biology pp. 29–46. In: Fifth Report of the West African Timber Borer Research Unit. Eyre and Spottiswoode, Chiswick, London.

Roberts, H. 1962b. Further observations on the biology of *Analeptes trifasciata* F. (Coleoptera: Lamiidae) with particular reference to the developmental stages pp. 81–90. In: Fifth Report of the West African Timber Borer Research Unit. Eyre and Spottiswoode, Chiswick, London.

Roberts, H. 1968. An outline of the biology of the mahogany shoot borer, *Hypsipyla robusta* Moore (Lep: Pyralidae) in Nigeria, with comments on other insect bark, stem and fruit borers of Nigerian Meliaceae. Commonwealth Forestry Review 47(30) NO. 133: 225–232.

Roberts, H. 1969. Forest Insects of Nigeria with Notes on their Biology and Distribution. Commonwealth Forestry Institute. Institute Paper 44, University of Oxford, 208 pp.

Rocheleau, D., F. Weber, and A. Field-Juma. 1988. Agroforestry in Dryland Africa. International Council for Research in Agroforestry. Nairobi, Kenya, 311 pp.

Ross, J.H. 1979. A conspectus of the Africa *Acacia* species, Chapter 4. Seed production predation, and dispersal, Memoirs of the Botanical Survey of South Africa, No. 44, 155 pp.

Sands, D.P.A. and S.T. Murphy. 2001. Prospects for biological control of *Hypsipyla* spp. with insect agents. In: *Hypsipyla* shoot borers in Meliaceae. Proceedings of an International Workshop held at Kandy, Sri Lanka, 20–23 August 1996. pp. 89–95.

Schabel, H.G. 2006. Forest Entomology in East Africa: Forest Insects of Tanzania. Springer, The Netherlands, 328 pp.

Shakacite, O. 1987. Seed problems in Zambia pp. 263–272. In: Kamra, S.K. and R.D. Ayling (eds) B. Proceedings of the International Symposium on Forest Seed Problems in Africa. IUFRO Report 7, 399 pp.

Speight, M.R. and D. Wainhouse. 1989. Ecology and Management of Forest Insects. Clarendon Press, Oxford, 374 pp.

Speight, M.R. and F.R. Wylie. 2001. Insect Pests in Tropical Forestry. CABI Publishing, New York, 307 pp.

Taylor, C.J. 1960. Synecology and Silviculture in Ghana. Thomas Nelson, London.

TEDB. 1987. Quarterly Report (January–March) of the Timber Export and Development Board, Ghana.

Tekulsky, M. 1985. The Butterfly Garden. The Harvard Common Press, Boston, 144 pp.

Thompson, G.H. 1959. *Trachyostus ghanaensis* Schedl., (Col.: Platypodidae), an ambrosia beetle attacking living wawa (*Triplochiton scleroxylon*) K. Schum. In: Ghana. Empire Forestry Review 38(4) No. 98: 420–421.

Thompson, G.H. 1960. Native farms and coleopoterous timber pests in Ghana. Empire Forestry Review 39(2) No. 100: 220–225.

Thompson, G.H. 1963. Forest Coleoptera of Ghana: Biological Notes and Host Trees. Oxford University Press, Amen House, London, 78 pp.

UNESCO. 1973. International Classification and Mapping of Vegetation. UNESCO, Paris.

Usher, M.B. 1978. Studies on a wood-feeding termite community in Ghana. West Africa. Biotropica 7(4): 217–233.

Usher, M.B. and J.K. Ocloo. 1979. The natural resistance of 85 West African hardwood timbers to attack by termites and microorganisms. Centre for Overseas Pest Research Tropical Pest Bulletin 6, 47 pp.

Wagner, M.R. and Cobbinah, J.R. 1993. Deforestation and sustainability in Ghana, West Africa. Journal of Forestry 91(6): 35–39.

Wagner, M.R., J.R. Cobbinah, and D.A. Ofori. 1996. Companion planting of the nitrogen-fixing *Gliricidium sepium* with the tropical timber species *Milicia excelsa* and its impact on the gall forming insect *Phytolyma lata* pp 264–271. In: Dynamics of Forest Herbivory: Quest for Patterns and Principles (eds) Mattson, W.J., Niemela, P., and M. Rousi. USDA Forest Service Technical Report, NC-183, 286 pp.

Wagner, M.R.V. Agyeman, J.R. Cobbinah, D.A. Ofori, and J.D. Nichols. 2000. Agroforestry and mixed species plantations as pest management strategies for *Phytolyma lata* in West Africa pp 65–78. In: Research Advances in Restoration of Iroko as a Commercial Species in West Africa (eds) Cobbinah, J.R. and M.R. Wagner. Proceedings of an International Workshop held in Kumasi, Ghana, 15–16 November 2000, 134 pp.

Wagner, M.R., S.K.N. Atuahene, and J.R. Cobbinah. 1991. Forest Entomology in West Tropical Africa: Forest Insects of Ghana. Kluwer Academic Publishers, Dordrecht, The Netherlands, 210 pp.

Watt, A.D., A.C. Newton, and J.P. Cornelius. 2001. Resistance in mahoganies to *Hypsipyla* species – A basis for integrated pest management. In: Hypsipyla Shoot Borers in Meliaceae. Proceedings of an International Workshop held at Kandy, Sri Lanka, 20–23 August 1996. pp. 89–95.

Watt, A.D., N.E. Stork, and M.D. Hunter (eds). 1997. Forests and Insects. Chapman & Hall, London, 406 pp.

White, M.G. 1964. The problem of the *Phytolyma* gall bug in the establishment of *Chlorophora*. Unversity of Oxford Commonwealth Forestry Institute Paper 37: 52.

White, M.G. 1968. Research in Nigeria on the Iroko gall bug (*Phytolyma* sp.) Nigerian Forest Information Bulletin 18: 72.

Williams, D.J. 1986. *Rastrococcus invadens* sp. N. (Hemiptera: Pseudococcidae) introduced from the Oriental Region to West Africa and causing damage to mango, citrus and other trees. Bulletin of Entomological Research 76: 695–699.

Williams, R.M.C. 1973. Evaluation of field and laboratory methods for testing termite resistance of timber and building materials in Ghana, with relevant biological studies. Centre for Overseas Pest Research Tropical Pest Bulletin 3: 67.

Willink, E. and D. Moore. 1988. Aspects of the biology of *Rastrococcus invadens* Williams (Hemiptera: Pseudococcidae), a pest of fruit of crops in West Africa, and one of its primary parasitoids, *Gyranusoidea tebygi* Noyes (Hemiptera: Encyrtidae). Bulletin of Entomological Research 78: 709–715.

Wong, J.L.G. 1989. Ghana Forest Inventory Project Seminar Proceedings 29–30 March 1989, Accra, Ghana. Forest Inventory Project, 101 pp.

Wylie, F.R. 2001. Control of *Hypsipyla* spp. Shoot borers with chemical pesticides; a review. In: Hypsipyla Shoot Borers in Meliaceae. Proceedings of an International Workshop held at Kandy, Sri Lanka, 20–23 August 1996, pp. 109–115.

Xerces Society. 1990.Butterfly Gardening: Creating Summer Magic in Your Garden. Sierra Club Books, San Francisco, 192 pp.

Appendix A

Forest tree species of Ghana

Scientific name	Family	Abbrev. I	Abbrev. II	Local name	Trade name	Comm. Status I	Comm. Status II
Afrosersalisia afzelii (Engl.) A. Chev.	Sapotaceae	Afaf		Bakunini		Class IV	2
Afzelia africana Sm.	Caesalpiniaceae	Afaf		Papao	Afzelia	Class III	1
Afzelia bella Harms. Var. *gracilior* Keay	Caesalpiniaceae	Afbe	Afz	Paopaonua		Class III	1
Aidia genipiflora (DC.) Dandy	Rubiaceae	Aige	Aid	Otwe-nsono		Class IV	3
Albizia adianthifolia (Schumach.) W.F. Wright	Mimosaceae	Alad	Ala	Pampena		Class III	2
Albizia ferruginea (Guill. & Perr.) Benth.	Mimosaceae	Alfe	Alf	Awiemfo-samina	Akbuzua	Class III	1
Albizia glaberrima (Schumach. & Thonn.) Benth.	Mimosaceae	Algl	Alg	Okro-sante/Okro-akoa		Class IV	2
Albizia zygia (Dc.) J.F. Macbr.	Mimosaceae	Alzy	Alz	Okro	Okuro	Class III	1
Alchornea floribunda Muell Arg.	Euphorbiaceae	Alfl		Gyamanini		Class IV	3
Allanblackia parviflora A. Chev.	Gluittiferae	Alpa		Sonkyi	Tallow Tree	Class III	3
Alsodeiopsis staudtii Engl.	Icacinaceae	Alst				Class IV	3
Alstonia boonei De Wild	Apocynaceae	Albo	Als	Sinro	Alstonia	Class IV	1
Amphimas pterocarpoides Harms.	Caesalpiniaceae	Ampt	Amp	Yaya	Yaya	Class IV	1
Androsiphonia adenostegia Stapf	Passifloraceae	Anad				Class IV	3
Angylocalyx oligophyllus (Bak.) Bak. F.	Papilionaceae	Anol				Class IV	3
Anogeissus leiocarpus (DC.) Guill.	Combretaceae	Anle	Ale	Kane	Kane	Class IV	3

Species	Family	Code		Local name		Class	No.
Anopyxis klaineana (Pierre) Engl.	Rhizophoraceae	Ankl	Ano	Kokotie	Kokoti	Class III	1
Anthonotha fragrans (Bak. F.) Exell & Hillcoat	Caesalpiniaceae	Anfr		Totoro-nini		Class IV	3
Anthonotha macrophylla P. Beauv.	Caesalpiniaceae	Anma		Totoro		Class IV	3
Antiaris toxicaria Leschenault	Moraceae	Anto		Kyenkyen	Antiaris	Class IIb	1
Antidesma laciniatum Muell. Arg.	Euphorbiaceae	Anla	Ala	Fotonini		Class IV	3
Antidesma membranaceum Muell. Arg.	Euphorbiaceae	Anme		Numanumagyma		Class IV	3
Aporrhiz urophylla Gilg.	Sapincleceae	Apur				Class IV	3
Aptandra zenkeri Engl.	Olacaceae	Apze	Apt	Ayemtu-dua		Class IV	3
Argomuellera macrophylla Pax.	Euphorbiaceae	Arma		Mprepre		Class IV	3
Aubrevillea kerstingii Pellegr./A.	Mimosaceae	Auke		Dahoma-nua		Class IV	3
Aulacocalyx jasminiflora Hook. f.	Rubiaceae	Auja	Auj	Ntweson		Class IV	3
Baphia nitida Lodd.	Papilionaceae	Bani	Ban	Odwen		Class IV	3
Baphia pubescens Hook. F.	Papilionaceae	Bapu	Bap	Odwen-kobiri		Class IV	3
Beilschmiedia mannii (Meisn.) Benth. & Hook. F	Lauraceae	Bema	Bem	Twianka		Class IV	3
Bequaertiodendron oblanceolatum (S. Moore) Heine & J.H. Hemsley	Sapotaceae	Beob		Nanfuro		Class IV	3
Berlinia confusa Hoyle	Caesalpiniaceae	Beco		Kwatafompaboa-nini		Class IV	1
Bersama abyssinica Fres.	Melianthaceae	Beab	Ba	Esonodokono		Class IV	3

(continued)

(continued)

Scientific name	Family	Abbrev. I	Abbrev. II	Local name	Trade name	Comm.	Comm. Status I	Comm. Status II
Blighia sapida K onig.	Sapindaceae	Blsa	Bls	Akye		Class IV		2
Blighia welwitschii (Hiern) Radlk.	Sapindaceae	Blwe	Blw	Akyekobiri		Class IV		2
Blighia unijugata Bak.	Sapindaceae	Blun		Akyebiri	Akyebiri	Class IV		2
Bombax brevicuspe Sprague	Bombacaceae	Bobr	Bbr	Onyina Koben	Bombax	Class IV		1
Bombax buonopozense P. Beauv.	Bombacaceae	Bobu	Bbu	Akonkodie/Akata		Class IV		1
Buchholzia coriacea Engl.	Capparaceae	Buco	Buc	Esonobese		Class IV		3
Bussea occidentalis Hutch.	Caesalpiniaceae	Buoc	Bus	Kotoprepre		Class IV		2
Callichilia subsessilis (Benth.) Stapf.	Apocynaceae	Casu				Class IV		3
Caloncoba echinata (Oliv.) Gilg.	Flacourtiaceae	Stet		Awiewuo-nua		Class IV		3
Caloncoba gilgiana (Sprague) Gilg.	Flacourtiaceae	Cagi	Cag	Awiewu		Class IV		3
Calpocalyx brevibracteatus Harms.	Mimosaceae	Cabr	Cab	Atowatere		Class IV		2
Canarium schweinfurthii Engl.	Burseraceae	Casc	Can	Bedi-wo-nua	Canarium	Class IV		1
Canthium vulgare (K. Schum.) Bullock	Rubiaceae	Cavu	Cvu	Ogya-pam-nini		Class IV		3
Carapa procera DC.	Meliaceae	Capr	Cap	Krabese/ Kwakuobese	Crabwood	Class IV		3
Carpolobia lutea G. Don.	Polygalaceae	Calu		Ofewa/Ofowa		Class IV		3
Cassipourea congoensis R. Br. Ex. DC.	Rhizophoraceae	Caco		Kokotenua		Class IV		3
Cassipourea hiotou Aubrev. & Pellegr.	Rhizophoraceae	Cahi				Class IV		3
Ceiba pentandra (Linn) Gaertn.	Bombacaceae	Cepe	Cpe	Onyina	Ceiba	Class IV		1

Celtis adolfi-friderici Engl.	Ulmaceae	Cead	Cfr	Esa-kusua	Celtis(a)	Class III	2
Celtis mildbraedii Engl.	Ulmaceae	Cemi	Cmi	Esa	Celtis(m)	Class III	1
Celtis wightii Planch.	Ulmaceae	Cewi		Prenprensa/Es-afufuo		Class IV	2
Celtis zenkeri Engl.	Ulmaceae	Ceze	Cez	Esa-koko	Celtis(z)	Class III	1
Chaetacme aristata Planch.	Ulmaceae	Char		Esono-ankaa		Class IV	3
Chassalia corallifera (A. Chev. Ex De wild)	Rubiaceae	Chco				Class IV	3
Chassalia kolly I(Schumach.) Hepper	Rubiaceae	Chko				Class IV	3
Chassalia subherbacea (Hiern) Hepper	Rubiaceae	Chsu				Class IV	3
Chazaliella sciadephora (Hiern) Petit & Verde	Rubiaceae	Chsc				Class Ia.	3
Childlowia sanguinea Hoyle	Caesalpiniaceae	Chsa		Ababima		Class IV	2
Chrysophyllum giganteum A. Chev.	Sapotaceae	Chgi	Cgi	Asonfena/Kunfena		Class IV	1
Chrysophyllum perpulchrum Mildbr. Ex Hutch & Dalz	Sapotaceae	Chpe	Cpe	Ataben		Class IV	2
Chrysophullum pruniforme Pierre ex Engl.	Sapotaceae	Chpr		Duatadwe		Class IV	2
Chrysophullum subnudum Bak.	Sapotaceae	Chsu	Csu	Adasema		Class IV	1
Chytranthus carneus Radlk.	Sapotaceae	Chear		Onibonanua		Class IV	3
Chytranthus cauliflorus (Hutch. & Dalz.) Wickens	Sapotaceae	Cheau				Class IV	3
Chytranthus macrobotrys (Gilg.) Exell & Mendonea	Sapotaceae	Chma	Cma	Tromwie		Class IV	3
Chytranthus verecundus H. Halle & Ake Assi	Sapotaceae	Chve				Class IV	3
Clausena anisata (Willd) Hook. F. ex Benth.	Rutaceae	Clan	Cla	Samanobi		Class IV	3

(continued)

(continued)

Scientific name	Family	Abbrev. I	Abbrev. II	Local name	Trade name	Comm. Status I	Comm. Status II
Cleidion gabonicum Baill.	Euphorbiaceae	Clga		Mpawuo		Class IV	3
Cleistopholis patens (Benth.) Engl. & Diels	Annonaceae	Cipa	Clp	Ngonenkyene	Ngo-ne-nkyene	Class IV	2
Coffea ebracteolata (Hiern.) Brenan	Rubiaceae	Coeb				Class IV	3
Coffea rupestris Hiern	Rubiaceae	Coru				Class IV	3
Coffea togoensis A. Chev.	Rubiaceae	Coto				Class IV	3
Cola caricifolia (G. Don) K. Schum	Sterculiaceae	Coca		Osonkurobia/Anseaya		Class IV	3
Cola chlamydantha K. Schum	Sterculiaceae	Coch	Coc	Tana-nfre		Class IV	3
Cola gigantean A. Chev.	Sterculiaceae	Cogi	Cog	Watapuo		Class IV	2
Cola heterophylla (P. Beauv.) Schott & Endl.	Sterculiaceae	Cohe				Class IV	3
Cola millenii K. Schum	Sterculiaceae	Comi	Cm	Anansedodowa	Dodowa	Class IV	3
Cola nitida (Vent.) Schott & Endl.	Sterculiaceae	Coni	Cn	Bcsc		Class IV	3
Cola reticulata A. Chev.	Sterculiaceae	Core				Class IV	3
Cola umbratilis Brenan & Keay	Sterculiaceae	Coum		Tananfrebere		Class IV	3
Copaifera salikounda Heckel	caesalpiniaceae	Cosa	Cs	Ohwendua	Copaifera		1
Corynanthe pachyceras K. Schum	Rubiaceae	Copa	Cop	Pamprama			2
Coula edulis Baill.	Olacaceae	Coed	Coe	Badwe	Coula		2
Craterspermum caudatum Hutch	Rubiaceae	Crca	Crc	Dua-dee		Class IV	3
Craterispermum cerinanthum Heim/	Rubiaceae	Crcc		Afra-ni-Afei		Class IV	3
Crotonogyne manniana Muell. Arg.	Euphorbiaceae	Crma				Class IV	3

Species	Family	Code	Code	Name	Name	Class	No.
Crudia gabonensis Pierre ex Harms.	Caesalpiniaceae	Crga		Samantacnini		Class IV	3
Cylicodiscus gabunensis Harms.	Mimosaceae	Cyga	Cyg	Denya	Okan	Class IV	1
Cynometra ananta Hutch. & Dalz.	Caesalpiniaceae	Cyan	Cya	Anata		Class IV	1
Cynometra megalophylla Harms.	Caesalpiniaceae	Cyme	Cym	Ananta-akoa		Class IV	3
Dacryodes klaineana (Pierre) H.J. Lam	Burseraceae	Dakl	Dak	Adwea	Adwea	Class IV	2
Daniellia ogea (Harms.) Rolfe ex Holl.	Caesalpiniaceae	Daog	Dao	Ehyedua	Ogea	Class IV	1
Dasylepis brevipedicellata Chipp.	Flacourtiaceae	Dabr		Astratoaduanini		Class IV	
Deinbollia grandifolia Hook. F.	Sapindaceae	Degr	Deg	Mmata		Class IV	3
Deinbollia pinnata (Poir) Schum. & Thonn.	Sapotaceae	Depi	Dep	Waagye-akoa		Class IV	3
Delpydora gracili A. Chev.	Annonaceae	Degr				Class IV	3
Dennettia tripetala Bak. F.	Tiliaceae	Detr				Class IV	3
Desplatzia chrysochlamys (Midbr. & Burret)	Tiliaceae	Dech	Dcc	Esonowisamfie		Class IV	3
Desplatzia subericarpa Bocg.	Caesalpiniaceae	Desu	Dcs	Esonowisatieberc	Class IV		3
Dialium aubrevillei Pellegr.	Caesalpiniaceae	Diau	Dia	Duabankye	Duabankye	Class IV	1
Dialium dinklagei Harms.	Caesalpiniaceae	Didi	Did	Dwedweedwe	Dalium	Class IV	3
Dialium guineense Wild.	Dichapetalaceae	Diaga	Dig	Asenaa		Class IV	3
Dichapetalum barteri Engl.	Dichapetalaceae	Diba		Akusakusaz		Class IV	3
Dichapetalum guineense (DC.) Keay	Thymeleaeaceae	Dicga	Dic	ESonowidee		Class IV	3
Dicranolepis grandiflora Engl.	Thymeleaeceae	Digr					3
Dicranolepis persei Cummins	Ebenaceae	Dipe					3

(continued)

(continued)

Scientific name	Family	Abbrev. I	Abbrev. II	Local name	Trade name	Comm. Status I	Comm. Status II
Diospyros abyssinica (Hiern) F. White	Ebenaceae	Diab		Gblitso(Ga)		Class IV	3
Diospyros canliculata De. Wild.	Ebenaceae	Dica		Otwaberlc		Class IV	3
Diospyros chevalieri De. Wild.	Ebenaceae	Dich					3
Diospyros cooperi (Hutch. & Dalz.) F. White	Ebenaceae	Dico		French/Atweaa-bere			3
Diospyros ferrea (Wild.) Bakh	Ebenaceae	Dife		Omenawa-nini			3
Diospyros gabunensis Gurke	Ebenaceae	Diga		Kusibiri		Class IV	3
Diospyros heudelotii Hiern	Ebenaceae	Dihc		Omenewabere		Class IV	3
Diospyros kamerunensis Gurke	Ebenaceae	Dika		Omenewa	African Ebony	Class IV	3
Diospyros mannii Hiern	Ebenaceae	Dima		Atweafufuo		Class IV	3
Diospyros mespiliformis Hochst. Eg A.DC.	Ebenaceae	Dime		Keke		Class IV	3
Diopyros monbuttensis Gurke	Ebenaceae	Dimo		Atwerenamti		Class IV	3
Diospyros piscatorial Gurke	Ebenanceae	Dipi		Otweto-kese		Class IV	
Diospyros sanza-minika A. Chev.	Ebenanceae	Disa	Dsm	Kusibiri/ Sanza-minike	Kusibiri/ Flint-bark	Class IV	1
Diospyros soubreana F. White	Ebenanceae	Diso		Otweto		Class IV	3
Diospyros vignei F. White	Ebenanceae	Divig		Omenawa-hoa		Class IV	3
Diospyros viridicans Hiern	Ebenanceae	Divir		Otwea		Class IV	3

Diphasia angolensis (Hiern) Verdoorn	Rutaceae	Dian		Amuduro		Class IV	3
Distemonanathus benthamianus aill.	Caesalpiniaceae	Dibe	Dis	Bonsamdua		Class IV	1
Dracaena arborea (Wild.) Link	Agavaceae	Drar	Dar	Ntomme	Ayan	Class IV	3
Dracaena camerooniana Bak.	Agavaceae	Drea	Dea	Nenkyema		Class IV	3
Dracaena elliotii Bak.	Agavaceae	Drel				Class IV	3
Dracaena fragrans (Linn.) Ker. Gawl.	Agavaceae	Drfr				Class IV	3
Dracaena manii Bak.	Agavaceae	Drma		Kesene		Class IV	3
Drypetes aframensis Hutch.	Euphorbiaceae	Draf		Duamako/ Opaha-nini		Class IV	3
Drypetes aubrevillei Le andri	Euphorbiaceae	Drau				Class IV	3
Drypetes aylmeri Hutch. & Dalz.	Euphorbiaceae	Dray		Opoahafufuo		Class IV	3
Drypetes chevalieri Beill.	Euphorbiaceae	Drch		Katrika		Class IV	3
Drypetes floribunda Muell. Arg.) Hutch.	Euphorbiaceae	Drfl	Drf	Bedibisa		Class IV	3
Drypetes gilgiana (Pax) Pax & K. Hoffm.	Euphorbiaceae	Drgi		Katrikeinin		Class IV	3
Drypetes ivorensis Hutch. & Dalz.	Euphorbiaceae	Driv		Opaha-bere		Class IV	3
Drypetes leonensis Pax	Euphorbiaceae	Drlc		Opaha-nua		Class IV	3
Drypetes parvifolia	Euphorbiaceae	Drpa		Katrikabere		Class IV	3
Drypetes pellegrinii Le andri	Euphorbiaceae	Drpe		Opahakokoo		Class IV	3
Drypetes principum (Muell. Arg.) Hutch.	Euphorbiaceae	Drpr		Opaha		Class IV	3
Drypetes singroboensis Ake Assi	Euphorbiaceae	Drsi		Opaha-akoa		Class IV	3

(continued)

(continued)

Scientific name	Family	Abbrev. I	Abbrev. II	Local name	Trade name	Comm. Class	Comm. Status I	Comm. Status II
Ehretia trachyphylla C.H. Wright	Horaginaceae	Ehtr	Eht	Okyini		Class IV		3
Elaeis guineensis Jacq.	Palmac	Elgu	Ela	Abe		Class IV		3
Elaeophorbia grandifolia (Haw.) Croizat	Euphorbiaceae	Elgr		Akani		Class IV		3
Enantia polycarpa (DC.) Engl. & Diels	Annonaceae	Enpo		Duasika	Duasika	Class IV		3
Entandrophragma angolense (Welw.) DC.	Meliaceae	Enan	Ea	Edinam	Gedu Nohor	Class Ia		1
Entandrophragma candollei Harms.	Meliaceae	Enca	Ecl	Penkwa-akoa	Omu	Class IIa		1
Entandrophragma cylindricum (Sprague) Sprague	Meliaceae	Ency	Ec	Penkwa	Sapele	Class Ia		1
Entandrophragma utile (Dawe & Sprague) Sprague	Meliaceae	Enut	Eu	Efoobrodedwo	Utile	Class Ia		1
Eriocoelum racemosum Bak.	Japindaceae	Erra	Err	Onibona-koko		Class IV		3
Erythroxylum emarginatum Thonn.	Erythroxylaceae	Erem	Ere	Bankyinini		Class IV		3
Euadenia eminens Hook. F.	Carpparaceae	Euem		Dinsimkoro		Class IV		3
Euclinia longiflora Salisb.	Rubiaceae	Eulo		Gyaneya		Class IV		3
Eugenia calophylloides DC.	Myrtaceae	Euca				Class IV		3
Eugenia obanensis Bak. f.	Myrtaceae	Euob				Class IV		3
Ficus exasperate Vahl	Moraceae	Ficx	Fic	Nyemkyerenini		Class IV		2
Funtumia africana (Benth.) Stapf	Apocynaceae	Fuaf	Fua	Okai		Class IV		3
Funtumia elastica (Preuss) Stapf	Apocynaceae	Fuel	Fuc	Fruntum		Class IV		3

Gaernera coopoeri Hutch. & M.B. Moss	Rubiaceae	Gaco				Class IV	3
Garcinia afzelii Engl.	Guttiferac	Gaaf	Gaa	Nsoko		Class IV	3
Garcinia epunctata Stapf	Guttiferac	Gaep		Nsokonua		Class IV	3
Garcinia gnetoides Hutch. & Dalz.	Guttiferac	Gagn		Tweapea-koa		Class IV	3
Garcinia smeathmannii (Planch & Triana) Oliv.	Guttiferac	Gasm		Bohwe/Tweapea-bere		Class IV	3
Glyphaea brevis (Spreng.) Monachino	Tiliaceae	Glbr		Foto		Class IV	3
Greenwayodendron oliveri (Engl.) Verde	Annonaceae	Grol		Duabiri		Class IV	3
Guarea cedrata (A. Chev.) Pellegr.	Meliaceae	Guce	Ge	Kwabohoro	Scented Guarea	Class IIa	1
Guarea thompsonii Sprague & Hutch.	Meliaceae	Guth	Gt	Kwabohoronii	Guarea (Black)	Class IIa	1
Guibourtia ehie (A. Chev.) J. Leonard	Caesalpiniaceae	Gueh	Gc	Hyeduanini/Anokyehyedua	Bubinga	Class IIb	1
Hannoa klaineana Ouerrec Engl.	Simaroubaceae	Hakl		Fotie/Hotrohotro	Hotrohotro	Class IV	2
Heisteria parvifolia Sm.	Olacaceae	Hepa		Sikayia		Class IV	3
Heritiera utilis/Tarrietia utilis Sprague	Sterculiaceae	Heut		Nyankom	Niangon	Class Ic	1
Hexalobus crispiflorus A. Rich.	Annonaceae	Hecr	Hex	Duabaha	Duabaha	Class IV	2
Hildegardia barteri (Mast.) Kosterm	Sterculiaceae	Hiba	Hib	Bronyadua/Akyerewaa		Class IV	3
Holarrhena floribunda (G. Don) Dur. & Schinz	Apocynaceae	Hofl	Hof	Scsc	Oscsc	Class IV	3
Holoptelea grandis (Hutch.) Mildbr.	Ulmaceae	Hogr	Hog	Nokwa	Nokwa	Class III	3

(continued)

(continued)

Scientific name	Family	Abbrev. I	Abbrev. II	Local name	Trade name	Comm. Status I	Comm. Status II
Homalium stipoulaceum Welw. Ex Mast.	Flacourtiaceae	Host				Class IV	3
Hunteria eburnean Pichon	Apocynaceae	Hucb				Class IV	3
Hymenostegia afzelii (Oliv.) Harms.	Caesalpiniaceae	Hyaf	Haf	Takorowa	Kouekoue	Class IV	3
Irvingia gabonensiis (Aubry-Lecomte ex O'Rorker) Baill	Irvingiaceae	Irga	Irg	Abese-buo	Abese-buo	Class IV	3
Isolona campanulata Engl. & Dicls	Annonaceae	Isca				Class IV	3
Isolona coooperi Hutch. & Dalz. Ex Cooper & Record	Annonaceae	Isco				Class IV	3
Isolona deightonii Keay	Annonaceae	Isde				Class IV	3
Ixora laxiflora Sm.		Ixla				Class IV	3
Khaya anthotheca (Welw.) C. DC.	Meliaceae	Khan	Ka	Kruben/Dubin-fufuo	Anthotheca	Class Ia	1
Khaya grandifoliola C. DC.	Meliaceae	Khgr	Kg	Kruba	Mahogany	Class Ia	1
Khaya ivorensis A. Chev.	Meliaceae	Khiv	Ki	Dubin	African Mahogany	Class Ia	1
Klainedoxa gabonensis Pierrc ex Engl.	Irvingiaceae	KIga	Kla	Kroma	Kroma	Class IV	1
Lannea nigritana (Sc. Elliot) Keavy	Anacardiaceae	Lani				Class IV	3
Lannea welwitchii (Hiern) Engl.	Anacardiaceae	Lawe	Law	Kumnini		Class IV	2
Lasianthus batangensis K. Schum.	Rubiaceae	Laba				Class IV	3

Scientific name	Family	Code	Code	Local name	Common name	Class	No.
Lasiodiscus mannii Hook. F.	Rhamnaceae	Lama	Lam	Adafa		Class IV	3
Lecaniodiscus capanioides Planch ex Benth.	Sapindaceae	Lecu	Lcc	Odwendera		Class IV	3
Leptaulus daphnoides Benth.	Icacinaceae	Leda	Led	Afena-akoa		Class IV	
Leptonychia pubescens Keay	Sterculiaceae	Lepu		Fotonua		Class IV	
Lophira alata Banks ex Gaertn. F.	Ochnaceae	Loal	Lop	Kaku	Ekki	Class IIa	1
Lovoa trichilioides Harms.	Meliaceae	Lotr	Low	Dubinbiri	African Walnut	Class Ib	1
Lychnodiscus dananensis Aubrev. & Pellegr.	Sapindaceae	Lyda		Sukye		Class IV	3
Lychnodiscus reticulates Radlk.	Sapindaceae	Lyre				Class IV	3
Maerua duchesnei (De Wild.) F. White	Capparaceae	Madu				Class IV	3
Maesobotrya bateri (Bail.) Hutch.	Euphorbiaceae	Maba		Apotrewa		Class IV	3
Majidea fosteri (Sprague) Radlk.	Sapindaceae	Mafo	Maf	Ankyewa		Class IV / Class IV	3
Malacantha alnifolia (Bak.) Pierre	Sapotaceae	Maal	Mal	Afram-sowa		Class IV	3
Mallotus oppositifolius (Geisel.) Muell. Arg.	Euphorbiaceae	Maop		Anyanyanforowa		Class IV	3
Mammea africana Sabine	Guttiferae	Maaf		Bompagya	African Appel	Class IV	1
Manilkara obovata (Sabine & G. don) J.H. Hemsely	Sapotaceae	Maob		Berekankum	African Pear-wood	Class IV / Class Ib	1
Mansonia altissima (A. Chev.) A. Chev.	Sterculiaceae	Maal	Man	Oprono	Mansonia	Class IV	1
Maranthes aubrevillei (Pellegr.) Prance	Chrysobalanaceae	Maau				Class IV	3
Maranthes glabra (Oliv.) Prance	Chrysobalanaceae	Magl		Afamnini		Class IV	3

(continued)

(continued)

Scientific name	Family	Abbrev. I	Abbrev. II	Local name	Trade name	Comm. Status I	Comm. Status II
Maranthes robusta (Oliv.) Prance	Chrysobalanaceae	Maro		Afambere		Class IV Class IV	2
Mareya micrantha (Benth.) Muell. Arg.	Euphorbiaceae	Mami	Mmi	Dubrafo		Class IV	3
Markhamia lutea (Benth.) IL. Schum.	Bignoniaceae	Malu		Abogyaneboobere		Class IV	3
Massularia acuminate (G. don) Bullock ex Hoyle	Rubiaceae	Maac		Fobe		Class IV	3
Memecylon afzelii G. Don.	Melastomataceae	Meaf		Otwe-anini		Class IV Class IV	3
Memecylon aylmeri Hutch. & Dalz.	Melastomataceae	Meay				Class IV	3
Memecylon blakeoides G. Don	Melastomataceae	Mebl				Class IV	3
Memecylon cinnamomoides G. Don	Melastomataceae	Meci				Class IV	3
Memecylon guineense Keay	Melastomataceae	Megu				Class IV Class IV	3
Memecylon lateriflorum (G. Don) Bremek.	Melastomataceae	Mela		Otwese		Class IV	3
Memecylon membranifolium Hook. F.	Melastomataceae	Meme		Otwesebere		Class IV	3
Memecylon normandii Jac.-Fel	Melastomataceae	Meno				Class IV	3
Microdesmis puberula Hook. F. ex Planch	Pandaceae	Mipu		Ofema		Class IV	3
Milicia excelsa (Welw.) Benth.	Moraceae	Chex	Chl	Odum	Iroko	Class IV	1
Milicia regia A. Chev.	Moraceae	Chre	Chr	Odum-nua		Class IV	1
Millettia rhodantha Bail.	Papilionaceae	Mirth	Mir	Tetetoa		Class IV	3

Species	Family			Local name	Trade name	Class	No.
Millettia thonningii (Schum. & Thonn.) Bak.	Papilionaceae	Mirth	Mit	Tsaatso		Class IV	3
Millettia zechiana Harms.	Papilionaceae	Mize	Miz	Afrafraha	Abura	Class IV	3
Itragyna ciliate Aubrev. & Pellegr.	Rubiaceae	Mici		Subaha		Class IV	3
Monodora myristica (Gaertn.) Dunal	Annonaceae	Momy	Mom	Awedeaba		Class IV	2
Monodora tenuifolia Benth.	Annonaceae	Mote	Mot	Motokurawadua		Class IV	3
Morus mesozygia Stapf	Moraceae	Mome	Mo	Wonton	Difou/ Wonto	Class III	2
Myrianthus arboreus P. Beauv.	Moraceae	Myar	Mya	Nyankuma		Class IV	3
Myrianths tibericus Rendle	Moraceae	Myli	Myl	Nyankumanini		Class IV	3
Napoleona vogelii Hook. & Planch	Lecythidaceae	Navo	Nav	Obua		Class IV	3
Nauclea diderrichii (De Wild. & Th. Dur.) Merrill	Rubiaceae	Nadi	Nau	Kusia	Opepe	Class Ia	1
Neostenanthera gabonensis (Engl. & Diels) Exell	Annonaceae	Nega				Class IV	3
Nesogordonia papaverifera (A. Chev.) R. Capuron	Sterculiaceae	Nepa	Nes	Danta	Danta	Class IIb	1
Newbouldia laevis (P. Beauv.) Seemann ex Bureau	Bignoniaceae	Nela	Nel	Sasamasa		Class IV	3
Newtonia duparguetiana (Bail.) Keay	Mimosaceae	Nedu	Nd	Adadabo		Class IV	3
Ochna staudtii Englo. & Gilg.	Ochnaceae	Ocst				Class IV	3
Octoknema borealis Hutch. & Dalz.	Olacaceae	Ocbo	Ob	Kwaasiwa/wis-uboni		Class IV	3
Octolobus spectabilis Welw.	Sterculiaceae	Ocsp		Afinafi		Class IV	3
Olax gambecola Baill.	Olacaceae	Olga				Class IV	3
Olax subscorpioidea Oliv.	Olacaceae	Olsu	Os	Ahouhenedua		Class IV	3

(continued)

(continued)

Scientific name	Family	Abbrev. I	Abbrev. II	Local name	Trade name	Comm. Status I	Comm. Status II
Ongokea gore (*Hua*) Pierre	Olacaceae	Ongo	Ong	Bodwe	Bodwe	Class IV	3
Ouratea affinis (Hook. F.) Engl.	Ochnaceae	Ouaf				Class IV	3
Ouratea amplectens (Stapf) Engl.	Ochnaceae	Ouam				Class IV	3
Ouratea calophylla (Hoo. F.) Engl.	Ochnaceae	Ouca	Oca	Opunini		Class IV Class IV	3
Ouratea duparquetiana (Baill.) Gilg.	Ochnaceae	Oudu				Class IV	3
Ouratea reticulate (P. Veauv.) Engl.	Ochnaceae	Oure		Anansedua		Class IV	33
Ouratea sulcata (Van Tiegh.) Keay	Ochnaceae	Ousu				Class IV	3
Oxyanthus formosus Hook. F. ex Planch	Rubiaceae	Oxfo	Oxf	Korantema		Class IV	3
Oxyanthus pallidus Hiern	Rubiaceae	Oxpa				Class IV	3
Oxanthus racemosus (Schum. & Thonn.) Keay	Rubiaceae	Oxra				Class IV	3
Oxyanthus speciesus DC.	Rubiaceae	Oxsp		Korantema		Class IV	3
Oxyanthus unilocularis Hiern	Rubiaceae	Oxun	Oxu			Class IV	3
Pachylpodanthium staudii Engl. & Diels	Annonaceae	Past	Pat	Kumdwie/Duawisa		Class IV	2
Pachystela brevipes (Bak.) Baill. ex Engl.	Sapotaceae	Pabr	Pab	Aframsua		Class IV Class IV	3
Pancovia bijuga Wild.	Sapindaceae	Pabi		Iguan ntohwe		Class IV	3
Pancovia sessiliflora Hutch. & Dalz.	Sapindaceae	Pase				Class IV	3

Species	Family	Code	Code2	Local name	Local name 2	Class	No.
Pancovia trubinata Radlk.	Sapindaceae	Patu				Class IV	3
Panda oleosa Pierre	Pandaceae	Paol	Pan	Kokorboba	Afam	Class IV	2
Parinari excelsa Sabine	Chrysobalanaceae	Paex		Afam	Asoma	Class IV	2
Parkia bicolor A. Chev.	Mimosaceae	Pabi	Pab	Asoma		Class IV	1
Pauridiantha sylvicola (Hutch. & Dalz.) Bremek.	Rubiaceae	Pasy				Class IV	3
Pavetta genidifolia Schumach.	Rubiaceae	Page				Class IV	3
Pavetta ixorifolia Bremek.	Rubiaceae	Paix				Class IV	3
Penianthus zenkeri (Engl.) Diels	Menispermaceae	Peze		Kramankote		Class IV	3
Pentaclethra macrophylla Benth.	Mimosaceae	Pema		Ataa		Class IV	2
Pentadesma butyracea Sabine	Guttiferae	Pebu		Abotosebie	Ataa	Class IV	3
Pericopsis elata (Harms.) van Meeuwen	Papilionaceae	Peel	Per	Kokroduo	Afromosia	Class Ib	1
Petersianthus macrocarpus (P. Beauv.) Liben	Lecythidaceae	Pema		Esia	Essia	Class IV	1
Phyllocosmus africanus (Hook. F.) Klotzch	Ixonanthaceae	Phaf		Akokorabeditoa		Class IV	2
Phyllocosmus sessiliflorus Oliv.	Ixonanthaceae	Phse				Class IV	3
Piptadeniastrum africanum (Hook. F.) Brenan	Mimosaceae	Piaf	Pip	Dahoma	Dahoma	Class IV	1
Piptostigma fasciculatum (De Wild.) Paiva	Annonaceae	Pifa		Duasika-fufuo		Class IV	3
Piptostigma fugax A. Chev. Ex Hutch. & Dalz.	Annonaceae	Pifu				Class IV	3
Placodiscus bancoensis Aubrev. & Pellegr.	Sapindaceae	Piba	Plb	Kafuosonini		Class IV	3

(continued)

(continued)

Scientific name	Family	Abbrev. I	Abbrev. II	Local name	Trade name	Comm. Status I	Comm. Status II
Placodiscus oblongifolius J.B. Hall	Sapindaceae	Plob				Class IV	3
Pleiocarpa mutica Benth.	Apocynaceae	Plmu	Plm	Onwen-ma		Class IV	3
Pouteria altissima (A. Chev.) Aubre v. & Pellegr.	Sapotaceae	Poal		Asamfenana-bere	Aningeria	Class IV	1
Pouteria aningeria (A. Chev.)	Sapotaceae	Poan		Asamfena-nini	Aningeria	Class IV	1
Protomagabaria macrophylla K. Schum.	Rubiaceae	Prma		Agyahere-nua		Class IV	2
Psilanthus mannii Hook. F.	Rubiaceae	Psma				Class IV	3
Psychotria ivorensis De Wild.	Rubiaceae	Psiv				Class IV	3
Pterygota bequaertii De Wild.	Sterculiaceae	Ptbe	Ptb	Kyereyebere		Class IV	3
Pterygota macrocarpa K. Schum.	Sterculiaceae	Ptma	Ptm	Kyereye	Ptergota/ Koto	Class IV	1
Pycnanthus angolensis (Welw.) Warb.	Myristicaceae	Pyan	Pye	Otie	Homba	Class IV	1
Pycnocoma cornuta Muell. Arg.	Euphorbiaceae	Pyco		Kafickafic		Class IV	3
Pycnocoma macrophylla Benth.	Euphorbiaceae	Pyma		Kafickafic		Class IV	3
Raphia hookeri Mann. & Wendl.	Palmac	Raho		Adobe		Class IV	3
Rauvolfia cumminsii Stapf	Apocynaceae	Racu	Rac			Class IV	3
Ricinodendron heudelotii (Bull.) Pierre ex Pax	Euphorbiaceae	Rihe		Wama	Erimado	Class IV	2

Species	Family	Code		Local name	Class	
Rinorea brachypetala (Turcz.) O. Ktze.	Violaceae	Ribra			Class IV	3
Rinorea breviracemosa Chipp	Violaceae	Ribre			Class IV	3
Rinorea dentata (P. Beauv.) O. Ktze.	Violaceae	Ride			Class IV	3
Rinorea ilicifolia (Welw. Ex Oliv.) O. Ktze.	Violaceae	Riil		Apose-nini	Class IV	3
Rinorea kibbiensis Chipp	Violaceae	Riki		Duaborobac	Class IV	3
Rinorea oblongifolia (C.H. Wright) Marquand ex Chipp	Violaceae	Riob		Mpawutuntum	Class IV	3
Rinorea prasina (Stapf) Chipp	Violaceae	Ripr			Class IV	3
Rinorea subintegrifolia (P. Beauv.) O. Ktze.	Violaceae	Risu			Class IV	3
Rinorea welwitchii (Oliv.) O. Ktze.	Violaceae	Riwe		Aposenini	Class IV	3
Rothmannia hispida (K. Cshum.) Fagerlind	Rubiaceae	Bobi		Tukabo	Class IV	3
Rothmannia longiflora Salisb.	Rubiaceae	rolo		Samankube	Class IV	3
Rothmannia urcelliformis (Hiern) Bullock ex Robyns	Rubiaceae	Rour			Class IV	3
Rothmannia whitfieldii (Lindl.) Dandy	Rubiaceae	Rowh		Sabobe	Class IV	3
Salacia pallescens Olvi.	Celastraceae	Sapa		Ntwea	Class IV	3
Salacia uregaensis R. Wilczek	Celastraceae	Saur			Class IV	3
Scaphopetalum amoenum A. Chev.	Sterculiaceae	Scam	Sea	Nsoto	Class IV	3
Scottellia klaineana Pierre	Flacourtiaceae	Sckl	Odoko	Tiabutuo	Class IV	2

(continued)

(continued)

Scientific name	Family	Abbrev. I	Abbrev. II	Local name	Trade name	Comm. Status I	Comm. Status II
Scytopetalum tieghemii (A. Chev.) Hutch. & Dalz.	Scytopetalaceae	Scti	Scy	Oprim		Class IV	3
Sloetiopsis usambarensis Engl.	Moraceae	Slus				Class IV	3
Soyauxia grandifolia Gilg. & Stapf	Medusandraceae	Sogr		Abotesima		Class IV	3
Soyauxia velutina Hutch. & Dalz.	Medusandraceae	Sove		Abotesimanua		Class IV	3
Sphenocentrum jollyanum Pierre	Menispermaceae	Spjo		Kroman-kote		Class IV	3
Spondias mombin L.	Anacardiaceae	Spmo		Atoa		Class IV	3
Sterculia oblonga Mast.	Sterculiaceae	Stob	Sto	Ohaa		Class IV	2
Sterculia rhinopetala K. Schum.	Sterculiaceae	Strh	Ste	Wawabima	Sterculia brown	Class IV	1
Sterculia tragacantha Lindl.	Sterculiaceae	Sttr	Stt	Sofo		Class IV	2
Stereospermum acuminatissimum K. Schum.	Bignoniaceae	Stac	Sta	Esonotokwakofuo		Class IV	2
Strephonema pseudocola A. Chev.	Combretaceae	Stps	Stp	Awuruku/Tutuaba		Class IV	3
Strombosia glaucescens Engl.	Olacaceae	Stgl	Str	Afina	Afina	Class IV Class IV	1
Syzygium guineense (Wild.) DC	Myrtaceae	Sygu				Class IV	3
Voacanga [Tabernaemontana] africans (Stapf) Pichon	Apocynaceae	Tach		ObonawaOfruma, Bedaa		Class IV	3
Tabernaemontana crassa Benth.	Apocynaceae	Tacr				Class IV	2

Tabernaemontana pachysiphon honStapf	Apocynaceae	Tapa				Class IV	2
Talbotiella gentii Hutch. & Greenway	Caesalpiniaceae	Tage		Takorowanua		Class IV Class Ib	2
Teclea verdoorniana Exell. & Mendonea	Rutaceae	Teve		Owebiribi		Class IV	
Terminalia ivorensis A. Chev.	Combretaceae	Teiv	Ti	Emire	Idigbo	Class IV	1
Terminalia superba IEngl. & Diels	Combretaceae	Tesu	Ts	Ofram	Afara	Class IV	1
Tetrapleura tetrapetera (Schum. & Thonn.) Taub	Mimosaceae	Tete	Tet	Prekese		Class Ia	3
Thecacoris stenopetala (Muell. Arg.) Muell. Arg.	Euphorbiaceae	Thst				Class IV	3
Tieghemella heckelii Pierre ex A. Chev.	Sapotaceae	Tihe	Tie	Baku	Makore	Class IV	1
Tricalysia pallens Hierns	Rubiaceae	Trpa	tpa	Turomdua	Class IV	Class IV	3
Tricalysia reflexa Hutch.	Rubiaceae	Trrel				Class IV	3
Tricalysia reticulata (Benth.) Hiern	Rubiaceae	Trret		Tanuro-kese		Class IV	3
Trichilia megalantha Harms.	Meliaceae	Trme				Class IV	3
Trichilia monadelpha Thonn.) J.J. De Wilde	Meliaceae	Trmo		Tanduro	Tanduro	Class IV	3
Trichilia prieuriana A. Juss	Meliaceae	Tripr	Trip	Kakadikuro		Class IV	2
Trichilia tessmannii Harms.	Meliaceae	Trte		Tanuronini		Class IV	2
Trichosypha arborea (A. Chev.) A. Chev.	Anacardiaceae	Trar	Tra	Anku		Class IV	3
Trilepisium madagascariense/Bosqueia angolensis DC	Moraceae	Trma		Okure	Okuri	Class IV	2

(continued)

(continued)

Scientific name	Family	Abbrev. I	Abbrev. II	Local name	Trade name	Comm. Status I	Comm. Status II
Triplochiton scleroxylon K. Schu.	Sterculiaceae	Trsc	Tri	Wawa	Obeche	Class Ib	1
Turraea heterophylla Sm.	Meliaceae	Tuhe		Ehunanyankwa/ Abugyentia		Class IV	3
Turraneanthus africanus (Welw. Ex. C. DC.) Pellegr.	Meliaceae	Tuaf	Tur	Apapaye/Wan-sawa	Avodire	Class IV	3
Uapaca guineensis Muell. Arg.	Euphorbiaceae	Uagu	Uag	Kuntan		Class IV	2
Uvariastrum pierreanum Engl.	Annonaceae	Uvpi		Otwe-ehi		Class IV	3
Uvariodendron angustifolium (Engl. & Diels) R.E. Frics	Annonaceae	Uvan		Bommofokwadu		Class IV Class IV	3
Uvariodendron calophyllum R.E. Fries	Annonaceae	Uvea		Esonokwadu-kokoo		Class IV	3
Uvariodendron occidentale Le Thomas	Annonaceae	Uvoc		Esonokwadu		Class IV	3
Uvariopsis globiflora Keay	Annonaceae	Uvgl		Asumpa-akoa		Class IV	3
Vitex ferruginea Schum. & Thonn.	Verbenaceae	Vife		Otwentorowa		Class IV	3
Vitex micrantha Gurke	Verbenaceae	Vimi		Otwentorowanini		Class IV	3
Xylia evansii Hutch.	Mimosaceae	Xyev		Samantawa/ Abo-bobima		Class IV	2
Xylopoia guintasii Engl. & Diels	Annonaceae	Xyqu	Xyq	Obaa/hentea-dua		Class IV	3

Exylopia staudtii Engl. & Diels	Annonaceae	Xyst	Xys	Dua-nan	Duanam	Class IV	3
Xylopia villosa Chipp	Annonaceae	Xyvi	Xyv	Obaafufuo Oyaa		Class IV	3
Zanthoxylum leprieurii Guill. & Perr.	Rutaceae	Zale				Class IV	2

LEGEND

Commercial Status I (Osafo 1970; Parren and de Graaf 1995)
Class I: Especially valuable species
Class II: Tree species of general utility
Class II: Tree species of possible future economic importance
Class IV: All other species
Commercial Status II (Wong 1989; Parren and de Graaf 1995)
Class 1: Species registered as having been exported from Ghana 1973–1988
Class 2: Species attaining 70 cm dbh and occuring at a frequency of more than 1/km and not presently exported
Class 3: All other species

REFERENCES

Osafo, E.D. 1970. The development of silvicultural techniques applied to natural forests in Ghana. FPRI Technical No. 13. Kumasi, Ghana.
Parren, M.P.E. and de Graaf, N.R. 1995. The quest for natural forest management in Ghana, Côte d'Ivoire and Liberia. Tropenbos Series 13, Wageningen, The Netherlands. 199 pages.
Wong, J.L.G. 1989. Ghana Forest Inventory Project seminar proceedings 29–30 March 1989, Accra, Ghana. Forest Inventory Project, 101 p.

Appendix B

Termite Species Recorded in Ghana

Kalotermitidae
    Cryptotermes havilandi (Sjöstedt)
    C. brevis (Walker)
    Neotermes aburiensis
Rhinotermitidae
  Coptotermitinae
    Coptotermes intermedius (Silvestri)
    C. sjostedti (Holmgren)
    C. reductus (Sjöstedt)
  Rhinotermitinae
    Schedorhinoterms putorius (Sjöstedt)
Termitidae
  Amitermitinae
    Amitermes evuncifer (Silvestri)
    A. stephensoni (Harris)
    A. crucifer
    A. spinifer (Silvestri)
    Anenteotermes polyscolus (Sands)
    Astalotermes quietus (Silvestri)
    Microcerotermes brachygnatus (Silvestri)
    M. fuscotibialis (Sjöstedt)
    Cephalotermes rectangularis (Sjöstedt)
    Microcerotermes parvalus (Sjöstedt)
  Macrotermitinae
    Ancistrotermes cavithorax (Sjöstedt)
    A. crucifer (Sjöstedt)
    A. guineensis (Silvestri)
    Macrotermes bellicosus (Smeathman)
    M. subhyalinus (Rambur)
    Microtermes subhyalinus (Silvestri)

M. aluco (Sjöstedt)
Anoplotermes spp
**Odontotermes fidens* (Sjöstedt)
Ancistrolermes amphidon
Macrotermes natalensis
**O. pauperans* (Silvestri)
O. badius (Haviland)
O. sudaneensis (Sjöstedt)
**Pseudacanthotermes militaris* (Hagen)
Nasutitermitinae
Leptomyxotermes doriae (Silvestri)
Nasutitermes arborum (Smeathman)
N. elegantulus (Sjöstedt)
**N. latifrons* (Sjöstedt)
Nasutitermes diabolus (Sjöstedt)
**N. luja*e (Wasmann)
Fulleritermes tenebricus (Silvestri)
Foramini termes spph
Trinervitermes geminatus (Wasmann)
T. occidentalis (Sjöstedt)
T. oeconomus (Tragardh)
T. togoensis (Sjöstedt)
T. trinervius (Rambur)
Termitinae
**Allognathotermes hypogeus* (Silvestri)
A. aburiensis (Sjöstedt)
**Basidentitermes mactus* (Sjöstedt)
**Cubitermes subcrenulatus* (Silvestri)
Euchilotermes tensus (Silvestri)
**Procubitermes aburiensis* (Sjöstedt)
Thoracotermes macrothorax (Sjöstedt)
B. potens
C. silvestri
C. gaigei (Emerson)
B. aurivillii (Sjöstedt)
**Pericapritermes urgens* (Silvestri)
**Ophiotermes grandilabius* (Merson)
Promirotermes holmgreni (Silvestri)
Termes hospes (Sjöstedt)

---

*Wood-feeding species

Appendix C

Non-Native Trees that have been Introduced into Ghana

(Adapted from the Forest Products Research Institute Information Bulletin, No. 5, 1972, 2 pp.)

Botanical name	Local name	Countries of origin
Azadirachta indica	Neem	India
Aucoumea klaineana	Okoume	Gabon, Congos
Araucaria angustifolia	Parana Pine	Brazil
Araucaria cunninghamii	Hoop Pine	Australia, Brazil
Araucaria hustenii	Hoop Pine	Australia
Araucaria bidwillii	Bunya Pine	Australia, South Africa
Acacia mearnsii	Black and Green Mattle	Australia
Adenanthera pavonina	Peacock Tree	India
Broussonetia papyrifera	Paper Mulberry	India
Bambusa arundinaceae	Thorny Bamboo	India
Cassia siamea	Cigar Boxtree	Southeast Asia
Cedrela odorata	West Indian Cedar	Mexico
Cedrela toona	Burma Cedar	Mexico
Casuarina equisetifolia	Whistling Pine, Beef-wood, Beech She-Oak	Australia, Malay & Archipelago
Callitris intra tropica	Cypress Pine	Australia
Cupressus lusitanica	Mexican Cypress, Cedar of Goa	Kenya, Tanganyika
Cupressus macrocarpa	Monterey Cypress	Australia
Cupressus glabra	Hard Cypress	Australia, USA
Cupressus sempervirens	Mediterranean Cypress	Australia
Cupressus arizonica	Arizona Cypress	Mexico, China
Cupressus funebris	Funeral Cypress, Weeping Cypress	Greece, Australia
Chamaecyparis lawsoniana	Port Oxford Cedar	Australia, Greece, Crete
Cassia fistula	Golden Shower	India, Burma, Ceylon
Cassia nodosa	Pink Cassia	Southeast Asia
Cassia spectabilis	Indian Almond	Tropical America

(continued)

(continued)

Botanical name	Local name	Countries of origin
Cinnamomum zeylanicum	Cinnamon	India, Ceylon
Calophyllum inophyllum	Alexandra Laurel	India
Crescentia cujete	Tree Calabash	Central America
Dalbergia latifolia	Bombay Blackwood	Indian Rosewood
Dalbergia sissoo	Shisham, Sissoo	India
Dendrocalamus strictus	Male Bamboo	India
Delonix regia	Flamboyante, Flame Tree	Madagascar
Erythrina phlebocarpa	Phlebocarpa	Australia
Eucalyptus torelliana	Cadagi White Gum	Australia
Eucalyptus alba	White Gum	Australia
Eucalyptus citriodora	Lemon-scented Gum	Australia
Eucalyptus hybrid cadambae	Cadambae	Australia
Eucalyptus grandis	Rose Gum	Australia
Eucalyptus pilularis	Black Butt	Australia
Eucalyptus tereticornis	Forest Red Gum	Australia
Eucalyptus saligna	Sydney Blue Gum, Saligna Gum	Australia
Eucalyptus punctata	Grey Gum	Australia
Eucalyptus camaldulensis	River Red Gum	Australia
Eucalyptus deglupta (nandiniana)	Mindanao Gum	Australia
Eucalyptus siderophloia	Broad-leaved (Red) Iron Bark	Australia
Eucalyptus robusta	Swamp Mahogany	Australia
Eucalyptus salmonophloia	Salmon Gum	Australia
Eucalyptus regnans	Mountain Ash, Giant Gum, Tasmanian Oak	Australia
Eucalyptus resinifera	Red Mahogany	Australia
Eucalyptus paniculata	Grey Iron Bark	Australia
Eucalyptus melliodora	Yellow Box, Honey Box	Australia
Eucalyptus intermedia	Pink Bloodwood	Australia
Eucalyptus globulus	Blue Gum	Australia
Eucalyptus gummifera	Red Bloodwood	Australia
Eucalyptus ciceziana	Gympic or Queensland Messmate	Australia
Eucalyptus siderosylon	Mugya, Red Iron Bark	Australia
Eucalyptus botryoides	Southern Mahogany	Australia
Eucalyptus scabra	White Stringy Bark	Australia
Eucalyptus propinqua	Small-fruited Grey Gum	Australia
Eucalyptus patens	Swan River or West Australian Black Butt	Australia
Eucalyptus maculata	Spotted Gum	Australia
Eugenia malaccensis	Malay Apple	Southeast Asia
Flindersia brayleyana	Silkwood	Australia
Gmelina arborea	Gmelina	India

(continued)

(continued)

Botanical name	Local name	Countries of origin
Hura crepitans	Sand-box Tree	Tropical America
Hevea brasiliensis	Pararubber	Brazil
Jacaranda mimosifolia	Blue Jacaranda	Tropical America
Lecythis zabucajo	Paradise Nut	Brazil to Costa Rica
Legerstoemia speciosa	Queen of Flowers	Eastern Tropics
Melia azadirachta	Azadirachta	India
Melia composita	Lunumidella	Ceylon
Millingtonia hortensis	Indian Cork Tree	India and Burma
Michelia champaca	Champac	India (Mimilaya)
Mimusops elengi		Asia
Ochroma lagopus	Balsa	Central South America
Pinus radiata	Monterey Pine	USA
Pinus caribaea	Caribbean Pine	British
Pinus taeda	Loblolly Pine	Batama Island, USA
Pinus elliotii var elliotii	Slash Pine	USA
Pinus elliotii var densa	South Florida Slash Pine	USA
Pinus merkusii	Tenasserim Pine	Australia
Pinus khaya	Khasia Pine	India, South Vietnam
Pinus occidentalis	Haitan Pine	India, Burma, Philippines
Peltophorum pterocarpum	Copper Pods	South Tropical Africa
Santalum album	Sandal Wood	India
Sterculia foetida	Foetid	Asia Minor, West Indies
Swietenia mahogany	Cuban, Spanish, or American	Florida, Central America
Swietenia macrophylla	Mahogany	America
Samanea saman	Rain Trees	Tropical America
Tectona grandis	Teak	India, Burma
Taxodium distichum	Swamp, B. Cypress	North America, Mexico
Taxodium ascendens	Swamp Cypress	North America, Mexico
Terminalia catappa	Indian Almond	M. Peninsula
Tabebuia rosea	Pink Poni, Rosy Poni	Trinidad

Appendix D

Insect pests of high value trees with potential use in plantations 1*, 2*

Host	Part of Plant Affected	Pest	Pest Order/Family	Damage Caused
Acacia spp.	Shoot	*Chilades elensis* (DEM)	Lepidoptera: Lycaenidae	Attack the shoot
		Planacoccoides njalensis (Laing)	Lepidoptera: Lycaenidae	
		Nyctipao walkeri Btlr	Lepidoptera: Noctuidae	
		Catopsilia florella F	Lepidoptera: Pieridae	
		Charama nilotica Hmps.	Lepidoptera: Noctuidae	Attack green fruit
		Ferrisiana virgata (Ckll)	Hemiptera: Coccidae	
		Bruchidius uberatus Fahreaus	Coleoptera: Bruchidae	Attack green/dry pods
	Leaf	*Euproctis fasciata*	Lepidoptera: Lymantriidae	Defoliator
		Aonidiella orientalis Newstead	Hemiptera: Diaspididae	
		Ceroplastes destructor Newstead	Hemiptera: Coccidae	
		Hemiberlesia lataniae Signoret	Hemiptera: Diaspididae	
	Shoot	*Coccus hesperidum*	Hemiptera: Coccidae	Sap feeder
	Flower	*Megalurothrips distalis*	Coleoptera: Bruchidae	Attack flower
	Seed	*Bruchidius uberatus*	Coleoptera: Bruchidae	Attack seeds
		Callosobruchus maculates	Coleoptera: Bruchidae	Attack seeds
Adansonia digitata		*Distaniella theobromae* (Dist)	Hemiptera: Miridae	Sap feeder
		Dysdercus superstitiosus (F)	Hemiptera: Pyrrhochloridae	
		Odontopus sexpunctatus Lap	Lepidoptera: Noctuidae	Defoliator
Afzelia africana		*Gonometa christyi* Sharpe	Lepidoptera: Lasiocampidae	Defoliator
		Pachymetana guttata Aur	Lepidoptera: Lasiocampidae	
		Achaea catella Gn	Lepidoptera: Noctuidae	
Albizia adianthifolia		*Enmonodia capensis* Herr. Schaf	Lepidoptera: Noctuidae	Defoliator
		Pericyma mendax Walk	Lepidoptera: Nymphalidae	
		Charaxes eupale (Drury)	Lepidoptera: Nymphalidae	
Albizia ferruginea		*Planococcoides njalensis* (Laing)	Hemiptera: Coccidae	Sap feeder
Albizia zygia	Leaf	*Diacrisia attrayi* Roths	Lepidoptera: Arctiidae	Defoliator
		Nudaurelia dione	Lepidoptera: Saturniidae	Defoliator

	Species	Order: Family	Note
Shoot	*Acridoschema isidori* Chevr	Coleoptera: Cerambycidae	Borer
	Symmerus tuberculatus (synonym)	Coleoptera: Platypodidae	Borer
	Xyloperthodes orthogonius Lesne	Coleoptera: Bostrichidae	Borer
	Xyleborus sharpie Hap	Coleoptera: Scolytidae	Attack living saplings
Stem	*Platypodidae chaetastus*	Coleoptera: Platypodidae	Breed in logs
	Doliopygus chapuisi	Coleoptera: Platypodidae	Borer
	Doliopygus perminutissimus	Coleoptera: Platypodidae	Borer
	Doliopygus serratus Strohm	Coleoptera: Platypodidae	Penetrate deeply in wood
	Doliopygus unispinosus Schedl	Coleoptera: Platypodidae	Breeds in fallen trees
	Platypus hintzi Schauf	Coleoptera: Platypodidae	
	Triozastus elongatus Schedl	Coleoptera: Platypodidae	Infest sawn boards
	Eccoptopterus sexspinosus Motsch	Coleoptera: Curculionidae	Wood borer
	Xyleborus badius Eichh	Coleoptera: Scolytidae	Wood borer
	Xyleborus mascarensis Eichh	Coleoptera: Scolytidae	Wood borer
	Xyleborus semiopacus Eichh	Coleoptera: Scolytidae	Infest cut boards
	Xylopertha crinitarsis	Coleoptera: Bostrichidae	Breeds in small stem
Other	*Argyrostagma niobe* Weym	Lepidoptera: Lymantriidae	
	Achaea lienardi (Boisd)	Lepidoptera: Noctuidae	
	Enmonodia capensis Herr.Schaf	Lepidoptera: Noctuidae	
	Entomogramma pardus Guen	Lepidoptera: Noctuidae	
	Ericeia sobria Wlk	Lepidoptera: Noctuidae	
	Pericyma mendax Wlk	Lepidoptera: Noctuidae	
	Polydesma umbricola	Lepidoptera: Noctuidae	
Albizia zygia Leaf/shoot	*Euphoresia maculiscutum* Fairm	Coleoptera: Melolonthidae	Defoliator
	Nadasi splendens Druce	Lepidoptera: Lasiocampidae	Defoliator
	Euproctis utilis Swin		Defoliator
	Anua ophiusa (Cramer)	Lepidoptera: Lymantriidae	Defoliator
	Planococcoides njalensis (Laing)	Lepidoptera: Noctuidae	Sap feeder
	Lygus neavsi	Hemiptera: Coccidae	Shoot borer

(continued)

(continued)

Host	Part of Plant Affected	Pest	Pest Order/Family	Damage Caused
		Achaea albifimbria	Hemiptera: Miridae	Shoot borer
		Parallelia spp.	Lepidoptera: Noctuidae	
			Lepidoptera: Noctuidae	
Alchornia cordifolia	Leaf	*Euphoresia maculiscutum* Fairm.	Coleoptera: Melolonthidae	Defoliator
		Nadiasa splendens Druce	Lepidoptera: Lasiocampidae	Defoliator
		Euproctis utilis Swin.		Defoliator
		Anua ophiusa Cramer	Lepidoptera: Lymantridae	Defoliator
			Lepidoptera: Noctuidae	
	Shoot	*Planococcoides njalensis* Laing	Hemiptera: Coccidae	Sap feeder
		Lygus neavsi	Hemiptera: Miridae	Shoot borer
		Achaea albifimbria	Lepidoptera: Noctuidae	Shoot borer
		Parallelia spp.	Lepidoptera: Noctuidae	
Antiaris africana	Shoot	*Triozamia lambourni* Newstead	Homoptera: Pyralidae	Attack the sap trees of all ages
		Xyleborus ferrugineus	Coleoptera: Scolytidae	Attack large branches
				Wood borer
	Stem	*Xylopertha crinitarsis*	Coleoptera: Bostrichidae	
		Apate monachus Fabr	Coleoptera: Bostrichidae	
		Doliopygus conradti Strohm	Coleoptera: Platypodidae	Infest on wood
		Platypus hintzi Schauf	Coleoptera: Platypodidae	Infest sawn board
		Platypus intermedius Strohm	Coleoptera: Platypodidae	Stem borer
		Trachyostus schaufussi Schedl	Coleoptera: Platypodidae	Stem borer
		Xyleborus camerunus Hag	Coleoptera: Scolytidae	Attack sapwood
		Xyleborus scabrior Schedl	Coleoptera: Scolytidae	Breed on stem
Azadirachta indica		*Aonidiella orientalis* Newstead	Hemiptera: Diaspididae	Mainly leaves
		Ceroplastes destructor Newstead	Hemiptera: Coccidae	Flowers, leaves, fruits
		Apate monachus Fabricius	Coleoptera: Bostrichidae	Stem borer
		Pseudaulacapis pentagona	Hemiptera: Diaspididae	

Blighia sapida	Shoot	*Premnobius cavipennis* Eichh	Coleoptera: Scolytidae	Bore fresh/dead branchwood
		Acridoschema isidori Chevr	Coleoptera: Cerambycidae	Sap feeder
		Doliopygus conradti Strohm	Coleoptera: Platypodidae	Wood borer
		Doliopygus perbrevis Schedl	Coleoptera: Platypodidae	Attack recently felled trees
		Doliopygus serratus Strohm	Coleoptera: Platypodidae	Breeds in logs/dying wood
		Platypus hintzi Schauf	Coleoptera: Platypodidae	Infest sawn board
Carapa procera	Fruit	*Ceratitis calae* Silv	Diptera: Trypetidae	Damage fruits
		Planococcoides njalensis (Laing)	Hemiptera: Coccidae	
Cedrela odorata	Leaves	*Godasa sidae*	Lepidoptera: Hispidae	Leaf skeletonizer
		Doliopygus conradti	Coleoptera: Platypodidae	
	Shoot	*Polygraphus busseae*	Coleoptera: Scolytidae	Breeds in branch wood
		Apate terebrans Pallens	Coleoptera: Bostrichidae	Insects bore into stem of young/healthy plant
Ceiba pentandra	Leaf	*Ascotis reciprocaria*	Lepidoptera: Geometridae	Defoliator
		Ascotis selenaria	Lepidoptera: Geometridae	Defoliator
	Shoot	*Analeptes trifasciata* Fabr	Coleoptera: Cerambycidae	Shoot borer
		Hypothenemus camerunus	Coleoptera: Scolytidae	Attack small branches
Ceiba pentandra	Stem	*Apate terebrans* Pallens	Coleoptera: Bostrichidae	Stem borer
		Dinoderus bifoveolatus	Coleoptera: Bostrichidae	
		Coccus hesperidum L	Homoptera: Coccidae	
		Crytophlebia leucotreta	Lepidoptera: Tortricidae	
		Distantiella theobromae (Distant)	Hemiptera: Miridae	
		Earias biplaga Walker	Lepidoptera: Noctuidae	
		Haritalodes derogaga Fabricius	Lepidoptera: Crambidae	
	Whole Plant	*Helopeltis schoutedeni* Reuter	Hemiptera: Miridae	
		Icerya seychellarum (Westwood)	Hemiptera: Margarodidae	
		Maconellicoccus hirsutus (Green)	Hemiptera:Pseudococcidae	
		Parasaissetia nigra (Nietner)	Hemiptera: Coccidae	

(continued)

(continued)

Host	Part of Plant Affected	Pest	Pest Order/Family	Damage Caused
		Planococcoides njalensis (Laing)	Hemiptera:Pseudococcidae	
		Planococcus citri (Risso)	Hemiptera:Pseudococcidae	
		Planococcus kenyae (Le Pelley)	Hemiptera:Pseudococcidae	
		Pseudaulacaspis pentagona (Targioni Tozzetti) MacGillivary	Hemiptera:Pseudococcidae	
		Pseudococcus longispinus Targioni Tozzetti	Hemiptera:Pseudococcidae	
		Selenothrips rubrocinctus (Giard)	Thysonoptera: Thripidae	
		Xylosandrus crassiusculus (Motschulsky)	Coleoptera: Scolytidae	
Celtis mildbraedii	Stem	*Acridoschema isidori* Chevr	Coleoptera: Cerambycidae	Sap feeder
		Trachyostus aterrimus Schauf	Coleoptera: Platypodidae	Sap feeder
		Doliopygus coelocephalus Schauf	Coleoptera: Platypodidae	Wood borer
		Doliopygus conradti Strohm	Coleoptera: Platypodidae	Infest logs
		Doliopygus dubius Samps	Coleoptera: Platypodidae	Breed in logs/dying trees
		Doliopygus exilis Chap	Coleoptera: Platypodidae	Attack felled trees
		Doliopygus perbrevis Schedl	Coleoptera: Platypodidae	Attack dry bark
		Doliopygus propinguus Schedl	Coleoptera: Platypodidae	Breed in logs/drying trees
		Doliopygus serratus Strohm	Coleoptera: Platypodidae	Breeds in fallen trees
		Doliopygus unispinosus	Coleoptera: Platypodidae	Attack felled trees
		Platypus hintzi Schauf	Coleoptera: Platypodidae	Insect of sawn timber
		Trachyostus aterrimus Schauf	Coleoptera: Platypodidae	Attack stems of living trees
		Trachyostus schaufussi	Coleoptera: Platypodidae	Attack logs with bark
		Hypothenemus socialis	Coleoptera: Scolytidae	Attack logs/bore into pith
		Xyleborus indicus	Coleoptera: Scolytidae	Confined to sapwood
		Premnobius cavipennis	Coleoptera: Scolytidae	
		Strombophorus ericius	Coleoptera: Platypodidae	Attack branch wood
		Xyleborus semiopacus Eichh	Coleoptera: Scolytidae	Breeds in small stems

Host / Plant part	Insect species	Order: Family	Habit/Damage
Shoot	Acmocera conjux Toms	Coleoptera: Cerambycidae	Phloem boring insect
	Acmocera compressa Fahr	Coleoptera: Cermbycidae	Phloem boring insect
	Apate monachus Fahr	Coleoptera: Bostrichidae	Bore into felled trees
	Dinoderus bifoveolatus	Coleoptera: Bostrichidae	Attack logs
	Xyleborus crinitarsis	Coleoptera: Scolytidae	Attack branch wood
	Xyleborus mascarensis Eichh	Coleoptera: Scolytidae	
***Entandrophragma* spp.**			
Entandrophragma angolense — Leaf	Nadasai splendens Druce	Lepidoptera:Lasiocampidae	Defoliator
Shoot	Hypsipyla robusta Moore	Lepidoptera: Pyralidae	Shoot borer
Fruit	Catopyla dysorphnaea	Lepidoptera: Pyralidae	Larvae feed on the seed
	Mussidia nigrivenella Ragonot	Lepidoptera: Pyralidae	
Entandrophragma angolense — Stem	Chaetastus tuberculatus	Coleoptera: Platypodidae	Breeds in logs
	Doliopygus conradti	Coleoptera: Platypodidae	Stem borer
	Doliopygus dubius	Coleoptera: Platypodidae	Breeds in logs
	Periommatus camerunus	Coleoptera: Platypodidae	Stem borer
	Platypus hintzi	Coleoptera: Platypodidae	Infest sawn timber
	Xyleborus camerunus	Coleoptera: Scolytidae	Stem borer
	Xyleborus mascarensis	Coleoptera: Scolytidae	Confined to sapwood
	Xyleborus semiopacus	Coleoptera: Scolytidae	Confined to sapwood
Entandrophragma cylindricum — Fruit	Catopyla dysorphnaea	Lepidoptera: Pyralidae	Larvae feed on seed
Shoot	Hypsipyla robusta Moore	Lepidoptera: Pyralidae	Larvae bore into seeds
	Gyroptera robertsi	Lepidoptera: Pyralidae	Sap feeder
	Heterobostrychus brunneus Murr	Coleoptera: Bostrichidae	Sap feeder
Stem	Hypsipyla robusta Moore	Lepidoptera: Pyralidae	Wood borer
	Chaetastus tuberculatus	Coleoptera: Platypodidae	Breeds in logs/stressed trees
	Doliopygus conradti	Coleoptera: Platypodidae	Stem borer
	Doliopygus dubius	Coleoptera: Platypodidae	Breeds in logs

(continued)

(continued)

Host	Part of Plant Affected	Pest	Pest Order/Family	Damage Caused
		Platypus hintzi	Coleoptera: Platypodidae	Infest sawn boards
		Trachyostus aterrimus	Coleoptera: Platypodidae	Attack stem of living trees
		Cordylomera spinicornis	Coleoptera: Cerambycidae	Phloem boring insect
***Eucalyptus* spp.**	Leaf	*Orgyia basali affinis* (Holland)	Lepidoptera: Lymantriidae	Defoliator
		Strepsicrathes rhothia Meyr.	Lepidoptera: Tortricidae	
		Spodoptera litoralis (Boisduval)	Lepidoptera: Noctuidae	Girdles bark
		Analeptes trifasciata Fabr.	Coleoptera: Cerambycidae	
		Anaemerus tomentosa Fabr.	Coleoptera: Curculionidae	Leaf roller/defoliator of seedlings
		Apate terebrans	Coleoptera: Bostrichidae	Borer
		Phymateus kazschi	Orthoptera: Acrididae	General defoliators in Northern Ghana. Infests recently felled trees
	Shoot	*Platypus lintzi* (Schaufuss)	Coleoptera: Platypodidae	
		Saissetia coffease Walker	Hemiptera: Coccidae	
		Parasaissetia nigra (Nieter)	Hemiptera: Coccidae	
		Xyleborus perforans (Wollaston)	Coleoptera: Scolytidae	
Gliricidia sepium	Stem	*Xylopertha crinitarsis*	Coleoptera: Bostrichidae	Attack felled trees
		Apate terebrans Pallens	Coleoptera: Bostrichidae	Stem borer
		Apate monachus Fahr	Coleoptera: Bostrichidae	Stem borer
Gmelina arborea	Leaf	*Diacrisia lutescens*	Lepidoptera: Arctiidae	Sap feeder
	Stem	*Hypothenemus eroditus* Westwood	Coleoptera: Scolytidae	Attack seedlings
		Hypothenemus pussilus Westwood	Coleoptera: Scolytidae	Attack fruits
Guarea cedrata	Fruits	*Balanogastris kolas* Desh.	Coleoptera: Curculionidae	Breed in green/dried fruit.
		Menechamus discrepans Faust	Lepidoptera: Pyralidae	Larvae feed on seed/fruit wall

Khaya spp.

Species	Part	Insect	Order: Family	Habit
Khaya grandifoliola	Fruit	*Catopyla dysorphnaea*	Lepidoptera: Pyralidae	Lay eggs in fruit
	Shoot	*Pseudophacopteron zimmermanni*	Homoptera: Psyllidae	Gall former
		Hypsipyla robusta Moore	Lepidoptera: Pyralidae	Shoot borer
	Stem	*Platypodidae chaetastus*	Coleoptera: Platypodidae	Breed in wood
		Apate monachus Fabr	Coleoptera: Bostrichidae	Tunnel into small stems
Khaya ivorensis	Fruits	*Catopyla dysorphnaea*	Lepidoptera: Pyralidae	Fruit feeder
	Shoot	*Polygraphus granulatus*	Coleoptera: Scolytidae	Breeds in branchwood
		Hypsipyla robusta Moore	Lepidoptera: Pyralidae	Shoot borer
		Xyleborus sharpie Hop	Coleoptera: Scolytidae	Attack living saplings
	Leaf	*Udinia faraguarsoni*	Hemiptera: Diaspididae	Suck sap from leaves
	Stem	*Bostrichoplites*	Coleoptera: Scolytidae	Wood borer
		Xyloborus perforans	Coleoptera: Scolytidae	Breed in newly cut logs
		Xyloborus semiopacus Eichoff	Coleoptera: Scolytidae	Breed in newly cut logs
		Xyloborus sharpie	Coleoptera: Scolytidae	Breed in newly cut logs
		Poecilips sannio Schauff	Coleoptera: Scolytidae	Stem borer
		Chaetastus tuberculatus Chap	Coleoptera: Scolytidae	Wood borer
		Doliopygus conradti Strohm	Coleoptera: Platypodidae	Wood borer
		Doliopygus dubius Samps	Coleoptera: Platypodidae	Wood borer
		Xyleborus mascarensis Eichh	Coleoptera: Scolytidae	Wood borer
		Xyleborus semiopacus Eichh	Coleoptera: Scolytidae	Wood borer
		Cordylomera spinicor	Coleoptera: Cerambycidae	Breed in stem
		Xylion sercurfier Lesne	Coleoptera: Bostrichidae	Stem borer
		Xylopertha crinitarsis	Coleoptera: Scolytidae	Attack felled trees
Khaya senegalensis	Fruit	*Cataphyla dysorphnaea*	Lepidoptera: Pyralidae	Fruit feeder
		Cryptoblabes quidiclla Mll	Lepidoptera: Pyralidae	Fruit feeder
	Shoot	*Hypsipyla robusta* Moore	Lepidoptera: Pyralidae	Shoot borer
Leucaena glauca	Leaf	*Syagrus* sp.	Coleoptera: Eumolpidae	Defoliator
		Ferrisiana virgata	Hemiptera: Pseudococcidae	

(continued)

(continued)

Host	Part of Plant Affected	Pest	Pest Order/Family	Damage Caused
Lophira alata	Leaf twig	*Planococcoides njalensis* (Laing)	Hemiptera: Coccidae	Leaf twig
Lovoa trichiolides	Shoot	*Cordylomera spinicornis* F.		Larvae tunnel under bark
Mansonia altissima	Leaf	*Godasa sidae* Fabricites	Lepidoptera: Arctiidae	Skeletonizer
	Stem	*Xyleborus mascarensis* Eichh	Coleoptera: Scolytidae	Confined to sapwood
		Xyleborus indicus Eichh	Coleoptera: Scolytidae	Attack living saplings
		Xyleborus camerunus Hag	Coleoptera: Scolytidae	Attack branches
		Xyleborus ambasiusculus Egg	Coleoptera: Scolytidae	Attack stems
Milicia spp. (*M. excelsa* & *M. regia*)	Shoot	*Phytolyma lata*	Homoptera: Psyllidae	Gall former
Musanga cercropioides	Leaf	*Phytolyma fusca*	Homoptera: Psyllidae	Gall former
		Stictococcus sjostedu	Homoptera: Coccidae	Defoliator
Nauclea diderrichii	Shoot	*Doliopygus conradti*	Coleoptera: Platypodidae	Attack branchwood
		Xyleborus camerunus	Coleoptera: Scolytidae	Attack simple branches
		Xyleborus indicus	Coleoptera: Scolytidae	Sap feeder
		Xyleborus mascarensis	Coleoptera: Scolytidae	Attack living saplings
		Orygmophora mediofoveata (Hampson)	Lepidoptera: Noctuidae	Shoot borer
	Stem	*Xyleborus ambasius*	Coleoptera: Scolytidae	Stem borer
		Xyleborus ambasiusculus Egg	Coleoptera: Scolytidae	Stem borer
		Xyleborus semiopacus	Lepidoptera: Noctuidae	Stem borer
		Doliopygus v-grandis (Samps)	Coleoptera: Platypodidae	Stem borer
Nauclea latifolia		*Demarius splendidulus* (F)	Hemiptera: Coccidae	
	Leaf	*Anomis leona*	Lepidoptera: Noctuidae	Minor defoliator
Nesogordonia papaverifera	Leaf	*Ferrisiana virgata* (Ckll)	Hemiptera: Coccidae	
		Panococcus citri (Risso)	Hemiptera: Coccidae	
		Sahlbergella singularis Hagl	Hemiptera: Miridae	

Host	Part	Pest	Order: Family	Damage
Nesogordonia papaverifera	Stem	*Ferrisiana virgata* (Ckll)	Hemiptera: Coccidae	Stem borer
		Planococcus citri (Risso)	Hemiptera: Coccidae	Stem borer
		Sahlbergella singularis Hahl	Hemiptera: Coccidae	Stem borer
		Anomis leona (Schaus)	Lepidoptera: Noctuidae	Stem borer
Pericopsis elata	Leaf	*Lamprosema lateritialis*	Lepidoptera: Pyralidae	Defoliator/leaf roller
	Seed	*Laspeyresia tricentra*	Lepidoptera: Tortricidae	
Piptadeniastrum africanum		*Eudrapa labandodes* Hmps	Lepidoptera: Noctuidae	
		Pericyma mendax Wlk	Lepidoptera: Noctuidae	
Pycanthus macrophylla (Welw.) Warh	Leaf	*Planococcoides njalensis* (Laing)	Hemiptera: Coccoidae	Leaf
Senna (Casia) siamea		*Frankliniella schultzei* (Trybom)	Thysanoptera: Thrididae	
		Thrips tabaci Lindema	Thysanoptera: Thrididae	
		Maconellicoccus hirsutus (Green)	Hemiptera: Pseudococcidae	
		Planococcoides njalensis (Laing)	Hemiptera: Coccidae	
Sesbania grandiflora	Shoot	*Apate monachus*	Coleoptera: Bostrichidae	Borer
		Apate terebrans	Coleoptera: Bostrichidae	Borer
	Leaf	*Azygophleps scalaris* (syn. *Phragmateoeria scalaris*)	Lepidoptera: Cossidae	Defoliator
Sterculia rhinopetala K. Schum	Twigs	*Planococcoides njalensis* (Laing)	Hemiptera: Cossidae	Attack twigs
		Planococcus citri (Risso)	Hemiptera: Cossidae	
		Anomis leona (Schaus)	Lepidoptera: Noctuidae	
		Anomis micordonta Hmps.	Lepidoptera: Noctuidae	
	Stem	*Xyleborus* spp.	Coleoptera: Platypodidae	Breeds in logs/stressed trees
		Premnobius caripenni		
		Doliopygus serratus Strohmeyer	Coleoptera: Platypodidae	
		D. solidus	Coleoptera: Platypodidae	
Tectona grandis	Leaf	*Ascotis selenaria reciprocaria* (Walker)	Lepidoptera: Geometridae	Defoliator

(continued)

(continued)

Host	Part of Plant Affected	Pest	Pest Order/Family	Damage Caused
		Diacrisia investigatorum	Lepidoptera: Arctiidae	Defoliator (nursery)
		Orgyia basali Affinis	Lepidoptera: Lymantriidae	Defoliator
		Spenoptera littoralis (Boisduval)	Lepidoptera: Noctuidae	
		Zonocerus variegatus L	Orthoptera: Acrididae	Defoliator
	Shoot	Planococcoides njalensis (Laing)	Hemiptera: Coccidae	Sap feeder
	Stem	Analeptes trifasciata Fabr	Coleoptera: Bostrichidae	Gnaws the bark
		Analeptes terebrans Pallas	Coleoptera: Bostrichidae	Stem borer
		Hypothenemus eroditus Westwood	Coleoptera: Scolytidae	Attack seedlings/twigs
		Hypothenemus pussilus Westwood	Coleoptera: Scolytidae	Attack seedlings/twigs
		Doliopygus erichsoni Chap	Coleoptera: Platypodidae	Attack healthy trees
		Apate terebrans Pallens	Coleoptera: Bostrichidae	Stem borer
Terminalia ivorensis	Fruit	Apate kentzeni	Coleoptera: Bostrichidae	Causes premature fruit fall
	Leaf	Epicerura pulverulaenta	Lepidoptera: Notodontidae	Defoliator
		Epicerura perghiosa	Lepidoptera: Notodontidae	Defoliator
		Lechriolepis sp.	Lepidoptera:Lasiocampidae	Defoliator
		Torrix dinota	Lepidoptera: Tortricidae	Defoliator
		Trichotaphs spp.	Lepidoptera: Gelchiidae	Defoliator
	Shoot	Cryptoflata unipunctata	Homoptera: Flatidae	Nymph sucks sap
		Planococcoides njalensis	Hemiptera: Coccidae	Sap feeder
	Stem	Apate Bostrychopis Tonsa	Coleoptera: Bostrichidae	Bore into sapwood
		Doliopygus dubius Samps	Coleoptera: Platypodidae	Bore deep into wood
		Doliopygus serratus Strohm	Coleoptera: Platypodidae	Wood borer
Terminalia superba		Ferrisiana virgata (Ckll)	Hemiptera: Coccidae	
		Formicococcus togoensis Strickland	Hemiptera: Coccidae	
		Newsteadia wacri Strickland	Hemiptera: Coccidae	
		Planococcus citri (Risso)	Hemiptera: Coccidae	
			Lepidoptera: Noctuidae	

Plant	Part	Species	Order: Family	Damage
		Achaea lienardi (Boisd)	Hemiptera: Coccidae	Defoliator
		Coccus hesperidum L	Orthoptera: Acrididae	
		Zonocerus variegatus	Orthoptera: Acrididae	
Triplochiton scleroxylon	Fruit	*Apion ghanaensis* Voss	Coleoptera: Apionidae	Fruit borer
		Apion nithonomoides Voss	Coleoptera: Aionidae	Fruit borer
	Seed	*Characoma nilotica* Hamps	Lepidoptera: Noctuidae	Attack green fruits
		Chlyphipterix spp	Lepidoptera: Cosmopterigidae	Attack green fruits
		Selepa docilis	Lepidoptera: Noctuidae	Attack seeds
	Leaf	*Tortrix dinota*	Lepidoptera: Tortricidae	Defoliator
	Shoot	*Xyleborus ferrugineus* Fahr	Coleoptera: Platypodidae	Attack logs
		Hypothenemus camerunus	Coleoptera: Scolytidae	Attack small branches
		Hypothenemus socialis	Coleoptera: Scolytidae	Bore to pith
		Premnobius cavipennis	Coleoptera: Platypodidae	Attack branch wood
	Stem	*Diclidophloebia eastopi* Vondracek	Homoptera: Psyllidae	Weakens stem bearing seeds
		Doliopygus interjectus Chaetastus tuberculatus Chap	Coleoptera: Platypodidae	Wood borer
		Doliopygus conradti Strohm	Coleoptera: Platypodidae	Stem borer
		Doliopygus dubius Samps	Coleoptera: Platypodidae	Infest logs
		Xyleborus indicus Eichh	Coleoptera: Platypodidae	Breed in logs
		Platyscus auricomus Schedl	Coleoptera: Scolytidae	Confined to sapwood
		Xyleborus mascarensis Eichh	Coleoptera: Platypodidae	Attack small branches
		Xyleborus neogranulatus	Coleoptera: Scolytidae	Attack living sapling
		Xyleborus pseudoem Basius	Coleoptera: Scolytidae	Borer
		Xyleborus semiopacus Eichh	Coleoptera: Scolytidae	Borer
		Apate terebrans Pallens	Coleoptera: Scolytidae	Breed in small stems

(continued)

(continued)

Host	Part of Plant Affected	Pest	Pest Order/Family	Damage Caused
		Bostrychoplites cornutus Oliv	Coleoptera: Bostichidae	Borer
			Coleoptera: Bostrichidae	Attack sawn or seasoned timber
		Dinoderus bifoveolatus Woll	Coleoptera: Bostrichidae	
Vitellaria paradoxa	Leaf	*Cirina forda* Westwood	Lepidoptera: Saturniidae	Defoliator
		Bostra glaucalis Hmps		
	Fruits	*Nephoterix orphnanthes* Meys	Lepidoptera: Pyralidae	
		Mussidia nigrivenella Ragonot	Lepidoptera: Pyralidae	
	Shoots	*Cardiophorus quadriplagiatus* Er	Lepidoptera: Pyralidae	
		Xyloctonus scolytoides Eichh	Coleoptera: Elateridae	
		Curimosphera sengalensis Haag	Coleoptera: Scolytidae	Borer
		Glypus conspicuous Westwood	Coleoptera: Tenebrionidae	

1* Information synthesized from Forsyth 1966, CABI Forest Protection Compendium and this volume.
2* The list of insects does not include termites.

Glossary

Abdomen: the posterior of the three main body divisions, contains digestive and reproductive structures.

Alate: winged individual in social insects.

Anal lobe: a lobe in the posterior basal part of the wing.

Antenna: (pl. antennae) a pair of segmented appendages located on the head above the mouthparts and usually sensory in function.

Anterior: front, in front of. Anus: the posterior opening of the alimentary tract.

Apical: at the end, tip or outermost part.

Apodal: legless

Apterous: wingless

Asymmetrical: not alike on the two sides.

Axon: the process of a nerve cell which conducts impulses away from the cell body of the nerve.

Basal: at the base, near the point of attachment (of an appendage).

Beak: the protruding mouthpart structures of a sucking insect; proboscis.

Biotic potential: rate of reproduction, depending on the number of ova, broods, etc.

Bipectinate: having branches on two sides like the teeth of a comb.

Bisexual: with males and females.

Brood: the individuals that hatch from the eggs laid by one mother; individuals that hatch at about the same time and normally mature at about the same time.

Camouflage: deception, where an insect is made to look like something else or concealed altogether.

Cardo: basal segment of first maxillar.

Carnivorous: feeding on the flesh of other animals.

Caste: a form or type of adult in a social insect.

Caterpillar: an eruciform larva; the larvae of a butterfly, moth, sawfly or scorpionfly.

Caudal: pertaining to the tail or the posterior part of the body.

Cercus: (pl. cerci) one of a pair of appendages at the end of the abdomen.

Chitin: a nitrogenous polysaccharide occurring in the cuticle of arthropods.

Cirbarium: the sucking apparatus by means of which the food is drawn up through the proboscis and passed into the oesophagus.

Circumesophageal connectives: portion of ventral nerve cord that circumvents the esophagus.

Claspers: paired grasping organs at end of abdomen in many male insects; or the last pair of prolegs in caterpillars.

Clavate: club-like, or enlarged at the tip.

Clypeus: part of face between labrum and frons.

Compound eye: an eye composed of many individual elements or ommatidia, each of which is represented exactly by a facet; the external surface of such an eye consists of circular facets that are very close together, or of facets that are in contact and more or less hexagonal in shape.

Coxa: the basal segment of the leg.

Cradles: place in which early stages of an insect are nurtured.

Crepuscular: active or flying at dusk or dawn.

Crochets: hooked spines at the tip of the prolegs of lepidopterous larvae.

Crop: the dilated portion of the foregut just behind the esophagus.

Cross vein: a vein connecting adjacent longitudinal veins.

Cryptobiotic: hidden; animals whose habit is to always remain hidden.

Cuticle: a tough noncellular coating in many plants and animals.

Declivity: a downward slope.

Dehiscence: a splitting along a natural line.

Diapause: a condition of suspended animation; a resting period between stages of insect development, generally to avoid adverse environmental conditions.

Dimorphism: of two forms; as in sexual.

Distal: that part of a segment or appendage farthest from the body.

Diurnal: active by day.

Diverticula: branches of the esophagus in some members of Diptera and Hymenoptera.

Dorsal: top or uppermost; pertaining to the back or upper side.

Ectoparasite: a parasite that lives on the outside of its host.

Effector nerve: nerves which carry impulses outward from the central nervous system to the effector organs.

Ejaculatory duct: the terminal portion of the male sperm duct.

Elbowed antenna: an antenna with the first segment elongated and the remaining segments coming off the first at an angle.

Elytron: (pl. elytra) a thickened, leathery or horny front wing.

Endoparasite: a parasite that lives inside its host.

Entire: without teeth or notches, with a smooth outline.

Entomophagous: feeding on insects; applied to wasps that feed their young with larvae, etc.

Epicormic branching: extensive lateral bud development usually associated with terminal shoot damage.

Epidermis: the cellular layer of the body wall which secretes the cuticula.

Etiolation: becoming whitened or bleached from lack of sunlight.

Exarate: sulcated; sculptured.

Femur: (pl. femora) the third leg segment, located between the trochanter and the tibia.

Filament: a slender threadlike structure.

Filiform: hairlike or threadlike.

Fontanelle: a small, depressed, pale spot on the front of the head between the eyes (Isoptera).

Foregut: the anterior portion of the alimentary canal, from the mouth to the midgut.

Frass: solid insect excrement; also used to describe wood fragments made by wood-boring insects.

Frons: forehead; upper part of head, between vertex and clypeus.

Front: that portion of the face between the antenna, eyes and ocelli; the frons.

Furcula: the forked springing apparatus of the Collembola.

Fusiform: spindle-shaped.

Galea: the outer lobe of the maxilla, borne by the stipes.

Gallery: tunnels in wood caused by wood-boring insects, normally refers to tunnels associated with egg laying.

Ganglion: (pl. ganglia) a knotlike enlargement of a nerve, containing a coordinating mass of nerve cells.

Gastric caeca: fingerlike pouches extending both anteriorly and posteriorly from the stomach; for digesting food.

Genitalia: the sexual organs and associated structures; the external sexual organs.

Gill: evagination of the body wall or hindgut, functioning in gaseous exchange in an aquatic insect.

Gonads: organs producing reproductive cells.

Gregarious: living in groups.

Grub: a scarabaeoid form larva; a thick-bodied larva with a well-developed head and thoracic legs, without abdominal prolegs, and usually sluggish.

Habitat: the environment in which an organism lives.

Haltere: (or halter) a small knobbed structure on each side of the metathorax in Diptera representing the hind wings.

Head: the anterior body region, which bears the eyes, antenna and the mouthparts.

Hemimetabolous: having a simple metamorphosis like that in the Odonata, Ephemeroptera and Plecoptera.

Herbivorous: feeding on plants.

Hindgut: the posterior portion of the alimentary tract, between the midgut and anus.

Holometabolous: with complete metamorphosis.

Horny: thickened or hardened.

Host: the organism in or on which a parasite lives; the plant on which an insect feeds.

Humeral: pertaining to the shoulder; located in the anterior basal portion of the wing.

Humeral angle: the basal anterior angle or portion of the wing.

Hyperparasite: a parasite whose host is another parasite.

Hypopharynx: a median mouthpart structure arising postorally and anterior to the labium; the ducts form the salivary glands are usually associated with the hypopharynx.

Imago: the adult or reproductive stage of an insect.

Ingest: to take food into the digestive tract.

Instar: the stage of an insect between successive molts, the first instar being the stage between hatching and the first molt.

Interconnecting nerves: nerve cells which link sensory and effector nerves.

Joint: an articulation or two successive segments or parts.

Labellum: (pl. labella) the expanded tip of the labium.

Labial palpus: (pl. labial palpi) one of a pair of small feeler-like structures arising from the labium.

Labium: of the mouthpart structures; the lower lip.

Labrum: the upper lip, lying just below the clypeus.

Lacinia: the inner lobe of the maxilla, borne by the stipes.

Lamellate: with platelike structures or segments; lamellate antennae.

Larva: (pl. larvae) the immature stages, between the egg and pupa, of an insect having complete metamorphosis; the six-legged first instar of Acarina; an immature stage differing radically from the adult.

Lateral: on or pertaining to the side (i.e. the right or left side).

Linear: line-like, long and very narrow.

Longitudinal: lengthwise of the body or of an appendage.

Malphighian tubules: excretory tubes that arise near the anterior end of the hind gut and extend into the body cavity.

Mandibles: jaw; one of the anterior pair of the paired mouthpart structures.

Mandibulate: with jaws fitted for chewing.

Margined: with a sharp or keel-like lateral edge.

Maxilla: (pl. maxillae) one of the paired mouthpart structured immediately posterior to the mandibles.

Maxillary palpus: (pl. maxillary palpi) a small feeler-like structure arising from the maxilla.

Median: in the middle; lying along the midline of the body.

Membraneous: like a membrane; thin and more or less transparent (wings); thin and pliable (cuticula).

Mentum: the distal part of the labium, which bears the palpi and ligula.

Mesonotum: the dorsal sclerite of the mesothorax.

Mesosternum: the sternum, or ventral sclerite, of the mesothorax.

Mesothorax: the middle or second segment of the thorax.

Metamorphosis: change in form during development.

Metanotum: the dorsal sclerite of the metathorax.

Metasternum: the sternum, or ventral sclerite, of the metathorax.

Metathorax: the third posterior segment of the thorax.

Midgut: the mesenteron, or middle portion of the alimentary tract.

Molt: a process of shedding the exoskeleton; ecdysis; to shed the exoskeleton.

Monogamous: mating with only one female.

Monoecious: possessing both male and female sex organs, hermaphroditic.

Moniliform: beadlike, with rounded segments; moniliform antennae.

Morphology: the science of form or structure.

Mouth: anterior opening to the alimentary tract.

Nasute: an individual of a termite caste in which the head narrows anteriorly into a snoutlike projection.

Nocturnal: active at night.

Nodus: a strong cross vein near the middle of the costal border of the wing.

Nymph: an immature stage (following hatching) of an insect that does not have a pupal stage; the immature stages of Acarina that have eight legs.

Ocellus: (pl. ocelli) a simple eye of an insect or other arthropod.

Oesophagus: the narrow portion of the alimentary canal immediately posterior to the pharynx.

Oligopodous: possessing only three pair of thoracic legs.

Ommatidium: (pl. ommatidia) a single unit or visual section of a compound eye.

Omnivorous: eating any kind of food.

Ootheca: (pl. oothecae) the covering or case of an egg mass.

Organochlorine: pesticides in the chlorinated hydrocarbon group.

Organophosphorus: a phosphorus-containing organic pesticide that acts by inhibiting cholinesterase.

Ostium: (pl. ostia) a slitlike opening of the insect heart.

Ovary: the egg-producing organ of the female.

Ovate: egg shaped.

Oviduct: the tube leading away from the ovary though which the eggs pass.

Oviparous: egg laying.

Ovipositor: the egg-laying apparatus; the external genitalia of the female.

Palpus: (pl. palpi) a segmented process borne by the maxillae or labium.

Parasite: an animal that lives in or on the body of another living animal (its host), at least during a part of its life cycle.

Parenchyma: the soft unspecialized tissues of organs.

Parthenogenesis: reproducing by eggs that develop without being fertilized; virgin birth.

Pectinate: with branches or processes like the teeth of a comb; pectinate antennae.

Pharyngeal pump: apparatus for pumping food; found in Culicidae; similar to cicarium.

Pharynx: the anterior part of the foregut, between the mouth and the esophagus.

Pheromone: substances which are secreted to the outside by an individual and received by a second individual of the same species which causes a specific action.

Phytophagous: feeding upon plants.

Pilose: covered with hair.

Pinnate: feathery.

Plicate: folded.

Plumose: featherlike; plumose antennae.

Polyembryony: an egg developing into two or more embryos.

Polygamous: mating with many females.

Polypodous: bearing both thoracic and abdominal legs.

Posterior: hind or rear.

Predator: an animal that attacks and feeds on other animals, usually animals smaller or less powerful than itself.

Prementum: the distal part of the labium, distal of the labial suture, on which all the labial muscles have their insertions.

Proboscis: the extended beaklike mouthparts.

Proleg: one of the fleshy abdominal legs of certain insect larvae.

Pronotum: the dorsal sclerite of the prothorax.

Propodeum: the posterior portion of the thorax, which is actually the first abdominal segmented united with the thorax (Hymenoptera suborder Apocrita).

Prosternum: the sternum, or ventral sclerite, of the prothorax.

Prothorax: the anterior of the three thoracic segments.

Protrusible: cable of extension and retraction.

Proximal: nearer to the body, or to the base of an appendage.

Pterostigma: a thickened opaque spot along the costal margin of the wing, near the wing tip.

Pubescent: downy, covered with short fine hairs.

Pulvillus: (pl. pulvilli) a pad or lobe beneath each torsal claw.

Puncture: a tiny pit or depression.

Pupa: (pl. pupae) the stage between the larva and the adult in insects with complete metamorphosis, a nonfeeding and usually inactive stage.

Pupate: transform to a pupa.

Quiescent: not active, a state of hibernation.

Radius: the longitudinal vein between the subcosta and the media.

Raptorial: fitted for grasping prey; raptorial front legs of praying mantids.

Reclinate: inclined backward or upward.

Rectum: the posterior region of the hindgut.

Reticulate: like a network.

Rostrum: beak or snout.

Rudimentary: reduced in size, poorly developed, vestigial.

Sap-feeding insects: insects which feed on the fluid (sap) in a plant's vascular system.

Saprophyte: a plant living upon dead organic material and obtaining its food in soluble form.

Scale: microscopic units (feathers) on wings of Lepidoptera, on elytra of some weevils, on body of Thysansura and Collembola; or protective covering, puparium, or diaspid scale insects.

Scape: basal segment of the antenna.

Sclerite: a hardened body wall plate bounded by sutures or membraneous areas.

Sclerotin: protein found in the exoskeleton; formed from the linkage of anthropodin molecules; results in a rigid exocuticle.

Sclerotized: hardened.

Segment: a subdivision of the body or of an appendage; between joints or articulations.

Seminal receptacle: a receptacle for the temporary storage of sperm received from another organism during copulation.

Seminal vesicle: a structure, usually saclike, in which the seminal fluid of the male is stored before being discharged.

Sensory nerve: nerves which carry impulses inward from the sense organs to the central nerve system.

Serrate: toothed along the edge like a saw; serrate antennae.

Seta: (pl. setae) a bristle.

Setaceous: a bristlelike; staceous antennae.

Setiferous: bristle bearing.

Siphon: specialized respiratory structure found on the posterior end of aquatic insect larvae.

Skeletonizers: any of a variety of lepidopterous larvae that consume the parenchyma of leaves, reducing them to a skeleton of veins.

Spermatheca: the saclike structure in the female in which sperm from the male are received and often stored.

Spine: a thornlike outgrowth of the cuticule.

Spiracle: an external opening of the tracheal system; a breathing pore.

Spur: a moveable spine; when on a leg segment usually located at the apex of the segment.

Stellate: star-shaped; star-shaped structure.

Sternite: a sclerite on the ventral side of the body; the ventral sclerite of an abdominal segment.

Stigma: a thickening of the wing membrane along the costal border of the wing near the apex.

Stipes: (pl. stipites) the second segment or division of maxilla, which bear the palpus, the galea, and the lacinia.

Stomach: digesting and absorbing organ just posterior to the crop.

Stria: (pl. striae) a groove or depressed line.

Striate: with grooves or depressed lines.

Stridulate: to make a noise (chirp or creak) by rubbing two structures or surfaces together.

Style: a bristlelike process at the apex of the antenna.

Stylet: a needlelike structure; one of the piercing structures in sucking mouthparts.

Stylus: a short, slender, fingerlike process.

Subapical: located just proximal of the apex.

Subbrain: ganglionic center in the thorax region which controls the locomotory organs.

Subesophageal ganglion: the knotlike swelling at the anterior end of the ventral nerve cord, usually just below the esophagus.

Submentum: the basal part of the labium.

Suctorial: adapted for sucking.

Sulcate: with groove or furrow.

Suranal plate: chitinous plate on dorsum of last abdominal segment in some larvae.

Suture: an external linelike groove in the body wall or a narrow membraneous area between sclerites; the line of juncture of the elytra.

Symbiosis: the living together, in a more or less intimate association, of two species.

Symmetry: a definite pattern of body organization; bilateral symmetry, a type of body organization in which the various parts are arranged more or less symmetrically on either side of a media vertical plane; that is, where the right side and left side of the body are essentially similar.

Tarsal claw: a claw at the apex of the tarsus.

Tarsal pad: a flattened structure located under the tarsus; often secretes adhesives.

Tarsus: (pl. tarsi) that part of the leg beyond the tibia, consisting of one or more segments or subdivisions.

Taungya: agroforestry practice in which food crops are interplanted with forest trees.

Tegmen: (pl. tegmina) the thickened or leathery front wing of an orthopteran.

Telson: the posterior part of the last abdominal segment.

Tergite: a sclerite of the tergum; the dorsal surface of an abdominal segment.

Tergum: (pl. terga) the dorsal surface of any body segment.

Terminal: at the end; at the posterior end (of the abdomen); the last of a series.

Termitarium: termite nest.

Testis: (pl. testes) the sex organ in the male that produces sperm.

Thorax: the body region behind the head, which bears the legs and wings.

Tibia: the fourth segment of the leg, between the femur and the tarsus.

Trachea: (pl. tracheae) a tube of the respiratory system, line with taenidia, ending externally at a spiracle, and terminating internally in the tracheoles.

Tracheoles: the fine terminal branches of the respiratory tubes.

Transverse: across, at right angles to the longitudinal axis.

Trochanter: the second segment of the leg, between the coxa and the femur.

Tympanum: (pl. typana) a vibrating membrane; an auditory membrane or eardrum.

Vagina: the terminal portion of the female reproductive system, which opens to the outside.

Vas deferens: (pl. vas deferentia) the sperm duct leading away from the testis.

Vein: a thickened line in a wing.

Venation: system of veins of a wing.

Ventral: lower or underneath; pertaining to the underside of the body.

Vertex: the top of the head, between the eyes and anterior to the occipital suture.

Vesicle: a sac, bladder, or cyst, often extensible.

Vestigial: small, poorly developed, degenerate, nonfunctional.

Viviparous: giving birth to living young, not egg laying.

Wanting: absent.